SYSTEM BEHAVIOR
AND
SYSTEM MODELING

SYSTEM BEHAVIOR
AND
SYSTEM MODELING

Arthur A. Few
Department of Space Physics and Astronomy
Rice University
Houston, Texas

UNIVERSITY SCIENCE BOOKS
SAUSALITO, CALIFORNIA

University Science Books
55D Gate Five Road
Sausalito, CA 94965
Fax: (415) 332-5393

Managing Editor: Lucy Warner
Editor: Louise Carroll
NCAR Graphics Team: Justin Kitsutaka, Lee Fortier, Wil Garcia,
Barbara Mericle, David McNutt, and Michael Shibao
Cover Design: Irene Imfeld
Compositor: Archetype Typography, Berkeley, California

Third edition

Macintosh Version ISBN: 0-935702-88-1
Windows Version ISBN: 0-935702-83-0

Printed in the United States of America

10 9 8 7 6 5 4 3 2 1

A Note on the Global Change Instruction Program

This series has been designed by college professors to fill an urgent need for interdisciplinary materials on the emerging science of global change. These materials are aimed at undergraduate students not majoring in science. The modular materials can be integrated into a number of existing courses —in earth sciences, biology, physics, astronomy, chemistry, meteorology, and the social sciences. They are written to capture the interest of the student who has little grounding in math and the technical aspects of science but whose intellectual curiosity is piqued by concern for the environment. The material presented here should occupy about two weeks of classroom time.

For a complete list of modules available in the Global Change Instruction Program, contact University Science Books, Sausalito, California, fax (415) 332-5393. Information about the Global Change Instruction Program is also available on the World Wide Web at http://home.ucar.edu/ucargen/education/gcmod/contents.html.

Contents

Preface

Although the word modeling appears conspicuously in the title of this module, it is primarily concerned with system behavior. The module is based on paradigms of conceptual model building, constructing a system diagram to describe how one thinks a system behaves, and building a complete quantitative model, whether on paper or on a computer, to learn how systems work or behave. It introduces fundamental concepts such as conservation laws, reservoirs, flows, controls, feedbacks, time constants, drivers, steady state, exponential growth, and limits, and applies them to basic real-world examples. I believe that this module can serve as a primer to the language and concepts used throughout Earth system science. It can be used alone or with computer and software components.

For the learning environments in which computers can be employed to teach modeling, this module has the added feature of providing the introductory training to use the computer modeling software STELLA® II. The teaching examples in the module can be implemented and run by the students; these models can be expanded and modified to explore system behavior under different scenarios. New and more complex Earth system models can be created by the students as individual or group assignments. We have used this software at Rice University in five courses at a variety of levels; the simple models described in this module are appropriate for the general undergraduate student with little prior computer experience, although tutorial assistance in the computer lab will be necessary. The *Instructor's Manual* for this module contains some modeling problems for the advanced science students. High Performance Systems, Inc. (45 Lyme Road, Hanover, NH 03755), developers of STELLA® II software, have permitted us to include a demonstration version of STELLA® II, a tutorial training exercise, and a set of sample models. Students can use them to explore STELLA® II, to run the models in this module, to examine other models provided on the disk, and to create their own models. For more information on STELLA software, please fill out and return the enclosed reader service card.

The module is designed for students pursuing non-technical degrees and is not limited to those who have access to computers and STELLA® II software. Those who do not can simply skip over the modelers' sections circumscribed with ↓■ and ■↑ and return to the text for the system behavior descriptions. The concepts should be clear from the discussion and the modeling outcomes from the graphs provided in the text. The 15 exercises integrated into the text do not require a computer and most are appropriate for nonscience majors. Students are guided to the solutions of the more difficult exercises. The sidebars provide more technical material for those who are interested.

The Global Change Instruction Program at NCAR would appreciate learning about your experiences with this and other modules in this series. Contact the GCIP, c/o Barbara McDonald, Advanced Study Program, National Center for Atmospheric Research, P.O. Box 3000, Boulder, CO 80307.

The Instructor's Manual with problem solutions, modeling projects, and visual aids may be obtained from University Sciences Books, fax (415) 332-5393.

Arthur A. Few
Rice University

Acknowledgments

This instructional module has been produced by the the Global Change Instruction Program of the Advanced Study Program of the National Center for Atmospheric Research, with support from the National Science Foundation. Any opinions, findings, conclusions, or recommendations expressed in this publication are those of the author and do not necessarily reflect the views of the National Science Foundation.

Executive Editors: John W. Firor, John W. Winchester

Global Change Working Group

Louise Carroll, University Corporation for Atmospheric Research

Arthur A. Few, Rice University

John W. Firor, National Center for Atmospheric Research

David W. Fulker, University Corporation for Atmospheric Research

Judith Jacobsen, University of Denver

Lee Kump, Pennsylvania State University

Edward Laws, University of Hawaii

Nancy H. Marcus, Florida State University

Barbara McDonald, National Center for Atmospheric Research

Sharon E. Nicholson, Florida State University

J. Kenneth Osmond, Florida State University

Jozef Pacyna, Norwegian Institute for Air Research

William C. Parker, Florida State University

Glenn E. Shaw, University of Alaska

John L. Streete, Rhodes College

Stanley C. Tyler, University of California, Irvine

Lucy Warner, University Corporation for Atmospheric Research

John W. Winchester, Florida State University

This project was supported, in part, by the
National Science Foundation
Opinions expressed are those of the authors
and not necessarily those of the Foundation

INTRODUCTION
What Is a System?

System science is not a new idea, but it is receiving renewed attention today because many of the global problems facing humanity are complex ones that transcend the classical disciplinary boundaries between and within the natural and social sciences. System science provides a methodology for quantitatively describing the behavior of complex dynamic systems. Because of this, and because of the broad applicability of system science and the increasing numbers of global problems requiring interdisciplinary skills, system science will continue to increase in importance in all disciplines. The purpose of this Global Change Instruction Program (GCIP) module, *System Behavior and System Modeling,* is to introduce system behavior, system science methodology, and system modeling.

A system may be very simple, such as a bathtub full of water, or very complex, such as the Earth's climate system or the solar system. It may be entirely physical; it may be social, such as a political system; or it may include both human and physical components. Ultimately, the system under consideration in Earth system science is the entire universe; from this system we isolate and define a much smaller subsystem that we hope to understand.

The first step in defining a system is to identify its components and interactions, if any, with other systems. Some of the components may themselves be systems, making them subsystems of the larger system. If a system has no significant interactions with the outside universe, we call it an isolated system. The second step is to identify the interactions between the components within the system.

The process of defining a system can be approached on a qualitative or quantitative level. When we provide a quantitative description of a system we call it system modeling. The qualitative system description can also be very useful in identifying system components and interactions that are important to understanding and altering the system's behavior. Consider carbon dioxide in the Earth's atmosphere. The system will include atmospheric carbon dioxide, energy production from fossil fuels (which give off carbon dioxide when burned), and complex subsystems of human energy consumption, fossil fuel recovery and marketing, fossil fuel reserves, human cultural and sociopolitical factors, as well as the subsystem associated with conservation and development and marketing of alternate energy sources.

There are important interactions among the system components. Measurements reveal a steady annual increase in atmospheric carbon dioxide produced primarily by the burning of fossil fuels to produce energy. Total energy use depends upon two things: the human population and the per capita energy use. (In a quantitative system model we could break this down by nation or groups of nations with similar energy-use patterns.) The per capita energy use is influenced by lifestyle, income, fuel availability, fuel cost, and available alternatives. Lifestyle includes factors like personal transportation, house size, heating and cooling requirements, urban or rural environment, and conservation practices. This system that we have just defined is not an isolated system because we included only the carbon dioxide in the air, not the carbon in the oceans or living things, which absorb carbon dioxide from the atmosphere.

What happens if the population steadily increases and the per capita energy use remains constant? Atmospheric carbon dioxide continues its increase. What if the population is stabilized but the per capita income increases? Atmospheric carbon dioxide continues its increase. If we wish to stop the increase in atmospheric carbon dioxide, which of the system or subsystem components that we have identified can we realistically control? Population, fuel cost, available alternatives, and conservation practices are good choices.

In the long term one of the system components listed above will ultimately dominate the system behavior: fuel availability, because fossil fuel is a finite resource. But before we reach that point, the accumulating atmospheric carbon dioxide and its associated global warming may produce unwanted and harmful effects on the larger Earth system. In order to know what these effects may be, we need to be more detailed and complete in defining our system and quantitative in including the interactions within and between the systems. This will require a system model.

Exercise

In the discussion of qualitative modeling we briefly described a system relating human use of energy to the increase in atmospheric carbon dioxide. The purpose of this exercise is to build upon this system and examine the interactions in greater detail. Step-by-step procedures for completing it are below. The exercise does not have a unique correct answer, but your response should be internally consistent, reflect known system relationships, and include all of the important items and interactions.

Use an outline format to define the basic structure of this system; the major, first-level, outline items will list the important components and subsystems of the system, and the

next level of the outline will list the components of the subsystems. You may add components and subsystems beyond those discussed in the text, and you may even add subsystems to subsystems if you think it is necessary. You may use the simplified outline provided below or build your own system outline on it.

Now characterize each item in your outline as positive, + (increases in the item increase atmospheric carbon dioxide), or negative, – (increases in the item work to decrease atmospheric carbon dioxide). For example, "Human Population" is positive while "Fuel Taxes" is negative. You may find it helpful to add, delete, and redefine the items in your outline; if you have an acute shortage of negative items, you may need to add new items such as "Birth Control Practices" or "Energy Policy" at the appropriate place in the outline. In the simple system outline below these items are italicized to remind us that they are not fully in place and operational. Some major outline items will have both positive and negative subitems; in this case, indicate "+ or -" for the major outline item, or give it the sign of the most influential of its subitems.

Next show the interactions between the items on your outline with arrows. For example, you should have arrows from "Fossil Fuels" to "Atmospheric Carbon Dioxide" and from "Standard of Living" to "Per Capita Energy Use." All components of a subsystem implicitly interact with the subsystem itself; they need not be shown with arrows. Interactions from the hypothesized items should be shown with dashed-line arrows to indicate their provisional nature. As you complete this part of the exercise you may discover that there is a better sequence for your outline so that most of the arrows point up the outline to form a chain of interactions with "Atmospheric Carbon Dioxide" at the top. Try to show all of the interactions with a minimum number of arrows; you may want to eliminate the arrows without a clear purpose.

Now label each of the arrows with either an

"S" to indicate a strong interaction or a "W" to indicate a weak interaction. The "Fossil Fuels" to "Atmospheric Carbon Dioxide" link is strong because the energy production directly produces carbon dioxide, which is directly injected into the atmosphere. The "Influence and Persuasion" to "Conservation, Nuclear, or Alternative Energies" connection is weak because the interaction is voluntary and depends upon relative prices of energy and the capital investment required to convert to different energy sources.

Now trace the sequence of arrows leading from each of the negative items in your outline to "Atmospheric Carbon Dioxide" and characterize the strength of the complete connection

---

SIMPLIFIED SYSTEM OUTLINE FOR HUMAN INFLUENCE ON ATMOSPHERIC CARBON DIOXIDE

1. **Atmospheric Carbon Dioxide**
2. **Energy Production**
   - 2.1. Fossil Fuels
   - 2.2. Conservation, Nuclear, or Alternative Energies
3. **Human Energy Needs and Uses**
   - 3.1. Human Population
   - 3.2. Per Capita Energy Use
4. **Fossil Fuel Market = Price**
   - 4.1. Fuel Taxes
   - 4.2. Owned Reserves
   - 4.3. Imported Reserves
   - 4.4. Public Reserves
5. **Cultural, Social, and Political Influences**
   - 5.1. Standard of Living
   - 5.2. *Birth Control Practices*
   - 5.3. *Energy Policy*
   - 5.4. Influence and Persuasion

---

by the weakest link in the chain. Finally, list by priority (strength of the interaction chain) the items that can work toward reducing the increase in atmospheric carbon dioxide. Identify the high-priority items from the "Cultural, Social, and Political Influences" subsystem. How many strong interactions are currently active?

Discussion

We started with some vague ideas of how this system worked, and by imposing structure on them we have refined our understanding of the system. This activity probably confirmed some of our prior opinions and focused our thoughts on the interactions and the differences in the importance of various strategies in interacting systems. As we progressed from our initial, almost subjective, opinion of how this system works to a nearly quantitative diagram, we have gained confidence in our understanding of the system, and perhaps we have changed some of our opinions.

Imagine the next step in the process that we started above. Suppose that we assign a number between 1.0 and 0.0 to each of the interacting arrows in place of the "S" or "W," where 1.0 is the strongest interaction possible and 0.0 is no interaction at all. We can compute a number for each complete interaction chain by taking the product of all of the values in the chain. The resulting number represents the strength of the item in influencing the ultimate objective, such as reducing atmospheric carbon dioxide. This semi-quantitative approach assists in establishing priorities and in evaluating which, and how much, "negative" action is required to counteract a specific "positive" action. The final improvement in understanding the system is to make a dynamic model of it that can change with time so that the system can respond to changing conditions.

A few of the items in the outline require explanation. The fossil fuel reserves, 4.2–4.4, are separated into three types—owned, imported, and public—which correspond respectively to those that are owned by the energy producer or a private party who sells to the producer; imported by the energy producer; and public, such as those on nationally owned lands or offshore in national territory. We separate the three reserves because the energy producer uses a mix of them to keep the price of fossil fuel products low and because energy policy can interact with the three reserves in different ways. It should not be a surprise that if the only reserves available to an energy producer were the owned reserves, their value would quickly escalate.

I
What Is Modeling?

Conceptual Modeling

The mind forms a visual image of an object, system, or process.

Modeling is really something we do every day. We form a conceptual model of some object, system, or process by creating a mental image of it. The mental image is a model, and the activity of creating the mental image is modeling. Consider the word "atom." Your mind has probably conjured up a picture that is your model of an atom. Perhaps you see a central body made up of black and white spheres representing protons and neutrons packed closely together, and around this central body fly tiny black dots in orbits, the electrons. (Modern physics tells us that this is not the best description for an atom, but this simple picture is the starting model from which physicists build the more complex models of quantum physics.)

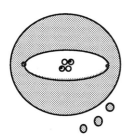

Communicating with Models

System behavior can be communicated with stylized drawings.

Have you noticed the widespread use today of icons, visual symbols often used to direct public behavior? A modern icon is a highly stylized model of an object or process (behavior). This international use of icons began when Volkswagen needed a universal language to communicate "headlights" and "windshield wipers" to the world's drivers. The use of graphics for communication has continued to expand as more products are marketed internationally. And, as more people travel internationally, there is an increasing need for highway and pedestrian signs that are language-free. In recent years we have seen the beginning of a revolution in the computer world, as icons replace the command languages that have for years dominated the human-computer interface. By using icons we can communicate mind-model to mind-model without translating our meaning through two languages.

The use of stylized drawings to depict the elements in a system and our understanding of how they interact has proven to be as valuable to the person modeling a system as the use of icons is to the international manufacturer. In addition to avoiding language problems, the use of system model diagrams is a more precise method of describing exactly what you, the model builder, have in mind.

Computer Models

Computers are used to model systems that are too complex to distill into a single statement or equation.

Creating numerical computer models of complex systems has become the most important application for large computers in science. In fact, it is problems like modeling the weather, climate, and location of fossil fuel reserves that are driving the computer industry to build larger and faster supercomputers. Although the supercomputers are necessary for complex, global models, smaller computers, even personal computers, also have an important role in the whole range of computer modeling.

What forces a model to require a large computer is usually spatial and temporal resolution. How detailed a picture, in space and time, does it provide? A weather forecast model that cannot portray the local weather in New York, Houston, and Seattle would have very limited usefulness; hence, a weather forecast model must have high spatial resolution. In an hour following sunrise, the surface temperature can increase several degrees. A weather forecast must have high temporal resolution to follow these changes or it cannot correctly portray the science that is occurring. For forecasting purposes, the Earth's surface is divided into many contiguous areas and the atmosphere above these areas is divided into layers; the volume elements formed by this process are called cells. The objective of weather and climate models is to forecast all of the atmospheric variables in all of the global cells for each specific time in the forecast period—the more times, the higher the temporal resolution.

Experience with large models has shown that the computational power requirements (computer memory size and arithmetic speed) are roughly proportional to the cube (power of three) of the number of cells in the model. Climate modelers would like to have global models with ten times better resolution than the models currently running on the world's most powerful computers; to fill this need, computing technology needs to be improved by a factor of a thousand!

Fortunately, for our purposes in this module we usually don't require high spatial or temporal resolution. In examining factors like global mean temperature and atmospheric CO_2 concentration, treating the Earth as a unit is adequate.

Exercises

1. The word model has many uses: a person that exhibits clothing, a miniature replica of an aircraft, exemplary behavior (as in a "model Boy Scout"), making three-dimensional objects from clay, etc. It has noun, adjective, and verb forms. Look up "model" in an unabridged dictionary to see the full breadth of this word. Why should it have such broad application? The answer is in the underlying meaning of the word: the visual image that is formed to represent the real object. Write two sentences employing "model" in each of its forms: noun, verb, and adjective (six sentences total). Write sentences that show the diverse uses of the word "model."

2. Suppose the size of a cell in a global model is to be 100 km by 100 km in the horizontal and divides the atmosphere into 15 layers. How many cells are required in the global model?

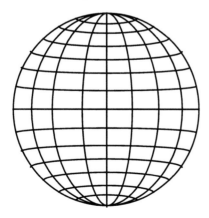

II

The Perspectives Provided by Modeling

Forecasting the Future

Models based upon understanding present system behavior are used to forecast the future.

Modeling is often used to forecast the future. This type of modeling occurs at every level of human activity from subconscious thinking to supercomputer modeling. You are actually using a conceptual model when you deal with the question, "Do I need to buy more milk?" Some of the procedures you use in responding to this question are (1) to check the level of the milk in the refrigerator; (2) to recall the rate at which the milk is being consumed in the household; (3) to predict how long the present supply will last; (4) to decide if the milk supply will last until the next shopping time. You have become so adept at modeling at this level that you probably have never thought about all of the steps that your mind takes to make this decision.

At the supercomputer level good examples of future forecast models are at the National Center for Atmospheric Research in Boulder, Colorado, and at the Goddard Institute for Space Studies in New York City. These groups use general circulation models (GCMs) to obtain moderate-resolution global forecasts of future climate conditions at the regional spatial level and the seasonal temporal level. GCMs provide detailed information including regional temperature, pressure, wind velocity, humidity, cloudiness, precipitation, and other derived parameters. (The GCMs are also called three-dimensional models because they treat latitude, longitude, and altitude as independent spatial variables.)

Most climate modelers agree that the climate models are not very accurate at the regional level because the cells used in them are typically 500 km on a side, about the area of the state of Colorado. This entire area must be represented by a single altitude, a single temperature, etc. In the real world there are large variations in all climate variables inside a cell of this size; there is no hope of correctly depicting regional conditions with such models.

The large computer models require a whole entourage of specialists to create and maintain the software and hardware, to research global conditions required to initiate model runs, and to interpret the model forecasts.

Between the extremely simple subconscious models and the very complex GCM climate models there is a complete range of modeling activities in which humans are continually engaged. Our ability to construct complex conceptual models for forecasting future situations may have been an important element in achieving evolutionary success. We can only hope that this same capability can prevent us from destroying the global environment.

Reconstructing Past Situations

Models are used to reconstruct past situations based upon remnant evidence of the past and an understanding of present system behavior.

We also use our modeling skills when we look backward in time. In your favorite detective or mystery novel, the main character's mission is probably "to reconstruct the crime." This is a conceptual modeling activity similar, but not identical, to the problem of forecasting the future. Both past and future models are dependent upon understanding present system behavior (the laws of nature applied to the system). We will see, however, that these two kinds of models require different methods.

Models of the future start with the conditions or parameters of the present, which we assume we know well; modelers refer to these as the initial conditions. For example, in our model of milk consumption, the amount of milk in the refrigerator was an initial condition. It is obvious that we must start the model with the correct values or we will never correctly forecast future values. Starting with the proper initial conditions, a future forecast model allows time to proceed forward and describes the status of all the system variables as a function of this progression in time. Some models evolve to a steady-state solution, which is independent of the initial conditions given to the model. The initial conditions determine the path taken by the evolving model, but the final state of the model can be reached by many different paths.

Why can't we just run a future forecast model backwards in time to describe some previous situation? The simple and straight-forward answer to this question is that nature doesn't work that way; we cannot force natural systems to run backwards.

When energy goes from one form to another, it always goes from a more concentrated, more useful form to a less concentrated, less useful one. For example, when a furnace burns oil or gas, the energy is dispersed into the atmosphere as heat. That energy is still present in the universe, but it cannot be recovered easily for human needs. Scientists use the concept of entropy to measure the disorder in a system. When concentrated energy is used to make dispersed energy, the system (which is the energy) is more disorganized; its entropy is greater. The second law of thermodynamics states that entropy always increases during natural processes.

It is the second law of thermodynamics that forces time to change in only one direction. Since entropy is one of the parameters that describe the state of a natural system, and since it can only increase, then the system itself can only change in such a manner that its entropy increases. Reversing time would require changes in the system that would decrease entropy; this is not allowed.

How, then, do we use a model to reconstruct the past? We must hypothesize a set of initial conditions for the system for a period of time that coincides with or just precedes the time that we want to model. The model is then run forward from that point in time to produce a set of system parameters for the time period being reconstructed. Why do you need the model if you know the initial conditions? Good question. The value gained in the modeling is that the model produces a complete set of consistent system descriptors, whereas the data from which the initial conditions were hypothesized were probably incomplete. For example, in modeling past Earth climates, geologists can tell us the location of the continents, the extent of the glaciers on the land, and the sea level for some particular time in the past. To this oceanographers can add the temperature of the sea surface, the volume of global ice, and the extent of sea ice. Climate modelers can use this information to hypothesize a set of initial conditions for a GCM; the GCM can then produce global data on air temperature, winds, precipitation, and other parameters that can be derived from the climate model variables. The model can tell us whether the glaciers were growing or shrinking, where specific plant and animal species could have been thriving, and the location of the major ocean currents.

How can the model start with a small quantity of input data and produce a complete global climate description? It has been pro-grammed with the laws of nature and has been tested (trained) to properly simulate the present Earth climate system; therefore, altering some of the initial conditions usually does not present the model with any problems that it cannot handle. If, however, the model is presented with initial conditions that require science that has not been included in the model, then the output will be incorrect, albeit probably interesting.

The GCMs are so large and require so much computer time to run that the normal operation for forecasting the future (starting with today's initial conditions and running forward to the desired future time) is often very difficult and expensive. For these large models the methodology for forecasting the future is the same as for reconstructing the past. A set of future initial conditions is hypothesized and the model is run to achieve a steady-state or equilibrium climate that is consistent with the hypothesized initial conditions. (Upon reaching the steady state, the average values of the system variables remain constant.) Since there are no records with which we can compare the output, there is no way to independently check the model's results. This technique is best used to evaluate deviations from today's climate owing to specified changes in the input parameters, such as the effect of doubling CO_2 in the atmosphere.

Sensitivity Studies
Models are used to evaluate a system's response to a specified change in one of its parameters or variables.

A sensitivity study examines how a model responds to a series of changes in its initial conditions or parameters. Frequently, the interaction being investigated involves only one component of the whole system. In such cases the sensitivity studies can be performed on a smaller model prior to involving a large, expensive GCM run.

Sensitivity studies are vital in evaluating the importance of various feedback components in the natural system. A feedback is a process that responds to a system change by enhancing or diminishing the change. (For example, if the Earth cools, ice sheets are likely to grow; they

will reflect more solar radiation, which causes further cooling. This process is a positive feedback.) The Earth system has many feedback processes, some of which we do not yet fully understand; sensitivity studies are useful in discovering which are more likely to produce global responses.

Another use for sensitivity studies is in investigating system behavior and interactions: how the system works. Unlike with forecast models, in which great care is taken to use the most accurate initial conditions and system parameters so that the model output will be believable, we can use sensitivity studies to push the system into unlikely situations that will expose the model to unanticipated responses and interactions. Sometimes this will expose errors in the model structure, and at other times it will provide new insight into the model dynamics. Sensitivity studies are tools for testing models, exploring model behavior, and learning how the system responds.

Understanding System Behavior
Modeling a system and running the model helps us understand how the system works.

In a rare case a newly created computer model will work on the first attempt. Experienced modelers and programmers can tell you many stories of "bugs" in their computer models that caused unexpected results and in some cases spectacular failures. These are almost never the computer's fault. Computers do exactly what they are told. Creating a good working model, even a simple one, forces the modeler to think clearly and succinctly about the system being modeled; computer models do not tolerate sloppy thinking.

Having created a working model, the next step is to test it. The testing is done by providing the model with input data for which there is a known result. One procedure for testing GCMs is to give them initial conditions that correspond to today's Earth (Earth-Sun relationships, location and size of continents and permanent ice, atmospheric gas concentrations, etc.) and allow the computer to run the GCM until the simulated climate system reaches a steady state solution. If the model forecasts ice sheets in Oklahoma, we should be suspicious. Other tests use extreme conditions. For example, if we turn the Sun off in the model, the Earth's temperature should evolve toward zero; if it doesn't, then we must go back to the drawing board.

When a model has been verified, we are ready to explore its performance and limits. This can be fun, like taking a new car out for a test drive. A range of system parameters and initial conditions is used to exercise all of the system interactions and to drive the system to certain defined limits. In addition to providing further testing of the model, this exploratory probing of its capabilities will reinforce our understanding of the system, and frequently the unexpected answers will give us new insights into system behavior.

Exercises

1. Treat the question "Can I afford to eat out tonight?" as we did the question "Do I need to buy more milk?" List all of the procedures and decisions that should go into responding to the question.

2. Consider a system composed of a marble and a wok and the system behavior when the marble is released at the edge of the wok. Describe the behavior of this system with different initial conditions, such as a simple release at the edge or a release with a sideways push on the marble. Does this system have a steady state solution? Discuss. Is the system behavior reversible with respect to time? Why?

III
The Components of Modeling

In Section I, we introduced the use of highly stylized visual models (icons) for communicating without language. In this section, we will use more formal diagram elements to communicate the structure, dynamics, and interactions of the systems we intend to model. Some of the disciplines using system diagrams are: electrical engineering, economics, computer programming, social sciences, business management, chemical engineering, civil engineering, and, of course, the Earth-related sciences. Within this wide range of disciplines that employ system diagrams, there is no universal set of symbols, but there are some basic similarities. The modeling skills that you learn here will be useful in many pursuits. Our symbol set for constructing system diagrams is rather generic. It was developed by Jay Forrester at the Massachusetts Institute of Technology and modified by High Performance Systems, Inc., for use with the STELLA® II software.

Reservoirs, Flows, and Valves
These are the fundamental elements in a dynamic system diagram. They define the variables of importance to the model and cause them to change with time.

Reservoirs are containers. Your mind model of a reservoir might be a water storage tank, an artificial lake, or a bathtub. The bathtub is a good working model for our purposes. The second element in this triad of system components is *flows*. In our bathtub model there are two obvious flows: in through the tap and out through the drain. It is important to note that flows are directional, adding water to the reservoir and draining water from it. The third element is *valves*. A valve controls a flow; the bathtub fill valve controls the rate at which water enters the bathtub, and the drain valve controls the rate at which water leaves the bathtub. The remaining system diagram components, *connectors* and *converters*, are introduced on pages 15 and 16; they process information for use by the system.

In constructing a system diagram, named rectangles are commonly used to represent reservoirs; heavy or double lines with arrows are used for flows, and circles or polygons attached to the flow symbol are used for the valves.

Reservoir   Flow and Valve

Because all flows have some form of valve, control, or restriction, and because a valve is useless without a flow to control, the flow-valve symbols in the system diagram are coupled as a single named symbol. The basic symbols representing reservoirs, flows, and valves are shown in the diagram above. (Names of the system diagram elements can be multiple words. When multiword names appear in equations, the blank spaces are underlined in equation lists to indicate that the compound name represents a single element of the system.)

The system diagram below represents our working model of bathtub dynamics.

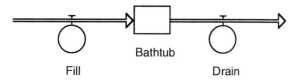

Reservoirs, flows, and valves sometimes go by other names in the many disciplines that use system diagrams. The word *stocks* is frequently used for *reservoirs*, *fluxes* for *flows*, and *controls* or *regulators* for *valves*.

Regardless of the names applied to these elements, there is a fundamental relationship between the reservoir (stock) and the flow (flux) that must be preserved.

All flows entering or leaving a reservoir must have the same units, and the units must represent the rate of change to the reservoir. If our reservoir represents liters of water in the bathtub, then the fill and drain flows must be given in liters per second. If the reservoir is your bank balance, the flows are then your monthly income and your monthly expenditures, all in dollars and cents. Or, if we consider population as a reservoir, the flows are annual births and annual deaths. The methodology of modeling can be applied to almost any problem, as long as we obey a small set of rules.

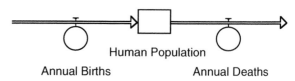

Another rule is that each reservoir must be provided with an *initial value*, the beginning value for the reservoir. The model then computes changes in the reservoirs during each model time step. In the population example, let us assume that the annual births and annual deaths are fixed; a forecast of population in the year 2000 is obviously dependent upon the population figures for the year that we initiate the model. If we err in the initial value, the model result will reflect that error.

We need to mention before leaving this topic that valves may be dynamic parameters themselves. In the population example, annual births and annual deaths are not fixed but are dependent upon population and other factors. In fact, it is the dynamic property of these parameters that makes them challenging to model.

Sources and Sinks

These are special reservoirs that represent the ultimate supply and repository for the dynamic reservoirs in the model. They are usually treated as large constant-level reservoirs relative to the system being modeled.

In the bathtub system diagram, we have fill water entering the bathtub and drain water leaving, but we have not specified the *source* of the fill water nor the *sink* into which the drain flows. To make the model logically complete, we need to add reservoirs onto both ends of the flow systems, as indicated in the figure below.

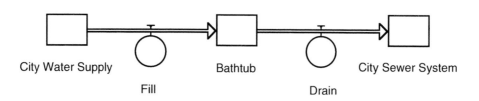

Reservoirs representing the primary resources from which flows originate in the model are sources; the "City Water Supply" is the source in this example. Reservoirs representing the final destination of flows in the model are sinks; in our example, the "City Sewer System" is the sink.

Sources and sinks are defined according to the problem being modeled. To the city engineer, who is not interested in the dynamics of one bathtub, the "City Water Supply" and the "City Sewer System" are the dynamic reservoirs in the model. She or he must look beyond these reservoirs to find sources and sinks for the whole city. The source might be wells, rivers, reservoirs, or, in the case of Boulder, Colorado, a glacier. The sink (after water treatment) is usually a river.

In many systems the sources and sinks are so large relative to the system being modeled that it is unnecessary to keep track of their levels; we simply represent each as an infinite "cloud," ⬡ .

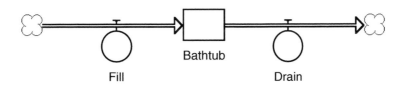

Bathtub
Fill   Drain

The best example of an infinite source and sink is the Earth's energy system itself. The solar radiation received by the Earth does not diminish the power put out by the Sun, and the infrared radiation emitted from the Earth into space does not modify space itself. The system diagram for the radiation part of the Earth's energy system is simple.

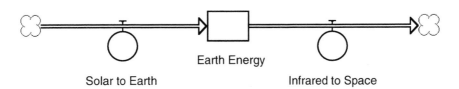

Earth Energy
Solar to Earth   Infrared to Space

There are other systems in which the source is finite and the system flows permanently diminish the reservoir. Examples are the exploitation of fossil fuels and tropical rain forests. And there are sinks that should be treated as finite reservoirs, such as garbage landfill sites.

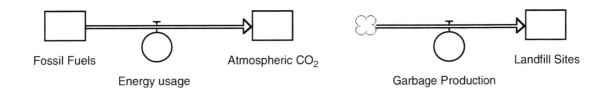

Fossil Fuels   Atmospheric CO_2 Landfill Sites
Energy usage   Garbage Production

There is also a special type of system that involves a closed loop between the source and sink; the closure may be complete, as in the case of the water supply on a long-duration space flight, or partial, as in paper recycling. The system diagrams for these two models are given below.

System Diagram for a Model of the Water Supply on a Long-Duration Space Flight

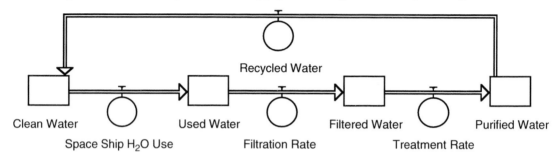

System Diagram for Paper Production Using Recycled Paper Pulp

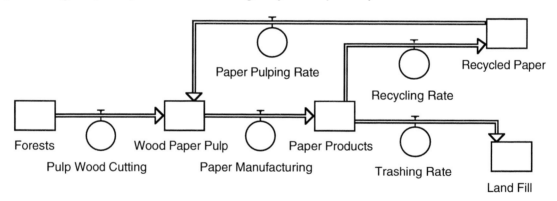

Branches and Decisions

In real systems, there are frequently multiple flows into and out of reservoirs; the modeler must provide the logic to apportion the flows among the available paths.

In the system diagram for recycling paper, we introduced a new concept. Did it bother you that "Wood Paper Pulp" had two input flows and "Paper Products" had two output flows? When a reservoir has multiple input or output flows we call them branches.

Having two input flows into a reservoir is a straightforward concept. When I drain my bathtub, the water flows through my pipes into the sewer system, and when you drain your bathtub, the water flows through your pipes into the sewer system; these actions are independent, and the sewer system accepts both input flows. Other branched input flows may require *decisions* from the model. For example the apportionment of wood pulp and paper pulp inputs into the "Wood Paper Pulp" reservoir will depend upon paper chemistry, economics, and availability of the two flows.

Having multiple input or output flows frequently requires providing the model with some sort of logic or set of rules so that the flow can be properly apportioned among the branches. The flows may be independent and the logic simple, or they may be linked or coupled, requiring more complicated logic. The logic incorporated into the model to apportion branched flows is a *decision* element of the system diagram. The decision element is usually incorporated into the valve component in the system diagram.

All bathtubs have overflow drains; they are usually hidden behind part of the fixture hardware and located approximately three quarters of the way up the side of the bathtub. The purpose, of course, is to allow water to overflow into the drain rather than onto the floor in the event that you carelessly fill the bathtub with too much water.

In the system diagram below we have included "DL" and "OL" in the valves to indicate the "Drain Logic" and the "Overflow Logic." We will not go into the logic itself here, except to note that the "Overflow" is zero until the bathtub reservoir level reaches the height of the overflow drain.

Interconnections and Coupling

Interconnections are used to pass information from one component of the system to another.

When one component of the system needs information from another component, we must provide an *interconnection* between them; such components are said to be *coupled*.

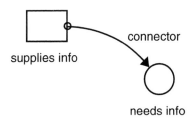

The symbol used in the system diagram to indicate the transfer of information between system components is a thin line with an arrow pointing in the direction of information transfer; it is different from the flow, which is a double line representing the flow of some material or property. In system diagram language the interconnection symbols are called connectors. Connectors are not named, and they must be attached on both ends.

In the bathtub overflow model discussed above we indicated that the overflow valve needed to know the level of the water in the bathtub in order for the "Overflow Logic" to take action. In the system diagram for this model we indicate the transfer of this information with a connector.

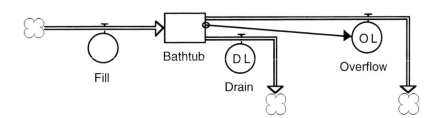

Inputs, Equations, and Decisions

A working model needs information on the system being modeled, in the forms of constants, parameters, variables, equations, and decision logic, in order to represent the relationships among all of the components in the system.

Thus far we have introduced *reservoirs, flows and valves,* and *connectors*; these three system diagram elements are the basic units used to build a system diagram for a model. We have also introduced *sources and sinks, branches and decisions,* and *interconnections and coupling,* which are special applications of the three basic elements. All of the basic behavior of your model is displayed on the system diagram by reservoirs, flows and valves, and connectors, but the model will not run until we provide it with additional information, such as inputs, equations, and decisions.

Converter

The fourth and final element used in the system diagram is the *converter*; the converter is a "catch all" system diagram element, used to represent all of the additional information required by the model. The computer programmer might refer to this function as a constant declaration or a subroutine; it is an important function of the model that is appended to the fundamental system logic. Like the valve, the converter is represented by a named circle, and the two elements can play very similar roles in the model; the valve, however, is always attached to a flow, whereas the converter is free-floating and can be placed anywhere in the system diagram.

One simple application for a converter is to supply the model with constants that it needs to compute other necessary parameters. In the

example model of the Earth energy system (page 13) we used a valve named "Solar to Earth." We assumed that somehow the valve would know how to compute the solar energy received by the Earth. The system diagram cannot read our minds or invent the necessary physics to compute "Solar to Earth." We must provide it with the values that it requires and the equation to use those values.

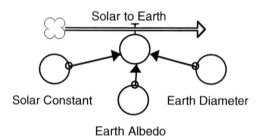

The solar constant is the solar radiant energy per square meter per second at the mean distance of the Earth from the Sun (1,368 Wm^{-2}). The albedo of the Earth is the percentage of sunlight (~30%) that it reflects back into space. The Earth's diameter (12.742 x 10^6 m) is needed to compute the cross sectional area of the Earth. Think of the solar radiation reaching the Earth at any given second as a plane, or sheet, in which the Earth makes an Earth-size hole. The size of that hole—the planet's cross section—is the total amount of solar radiation the Earth receives in that second.

These three constants are stored in converters and provided to the valve by the connectors, as shown above. The valve must now be instructed how to use the information provided to compute the solar energy received by the Earth system; this is done by giving the valve the equation in the box below.

In the equation below and those to follow the words and word strings with underlined spaces represent named components on the system diagram. Other conventions used in the

---

Solar to Earth Equation

Solar_to_Earth = Solar_Constant*(1-(Earth_Albedo/100))*PI*(Earth_Diameter/2)^2

equations are: the asterisk, *, for multiplication; the slash, /, for division; the caret, ^, for "raising to the power"; and PI for π = 3.14....

The equation can be written in a more familiar algebraic form E = S(1-A) πr^2, where E represents the solar energy absorbed by the Earth, S is the solar constant, A is the Earth's albedo expressed as a fraction, and πr^2 is the cross sectional area of the Earth. The two equations are the same mathematically. In the algebraic form we use single letters to represent the parameters and multiplication is implied by the adjacent placement of the letters, but we must define what each of the letters represents. In the computer version of the equation in the box we use descriptive names for the parameters in the equation, and all operations (such as multiplication: *) must be explicitly displayed.

The modeler could have put the values for the three constants directly into the equation rather than introducing them as converters then connecting them; this, however, is a bad practice because it does not explicitly show on the system diagram the identity and use of the parameters. Furthermore, if the modeler, in the course of testing or modifying the model, decides to change some parameter, it would be necessary to know every place in the model that the parameter had been used and change it in all of those places. When the parameter is shown as a converter, only one change is necessary.

Another frequent use for converters is the creation of new variables. Let us return to the Earth energy model system diagram (on page 13); this time we will direct our attention to the right side of the diagram. In order to compute "Infrared to Space," the Earth's radiation outward into space, we will need to know the temperature of the surface of the Earth. This, of course, is also what we want the model to tell us. One of the appealing aspects of dynamic modeling is that we can use a parameter as if we knew its value even though at the time we use it the value has not been computed.

In order to compute the temperature of the Earth's surface we need to use the concept of heat capacity from thermodynamics. When we heat an object, we know that the temperature of the object will increase. The same amount of heat will raise different objects to different temperatures. The property that determines its temperature change for a given amount of heat exchanged is the object's heat capacity: the thermal energy exchanged divided by the temperature change. For a given amount of added energy, an object with a large heat capacity will experience a small temperature increase, while an object with a small heat capacity will experience a larger increase.

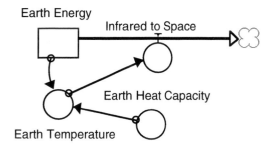

The temperature computation uses the first law of thermodynamics, which is also one form of the law of conservation of energy (see page 18). In our application here we compute the temperature of the Earth by dividing the energy content of the Earth surface, "Earth Energy," by the heat capacity of the Earth surface, "Earth Heat Capacity," a value that we must supply with another converter. The computed value for the temperature is then supplied to the valve "Infrared to Space" with a connector. We must also provide the converter "Earth Temperature" with the equation shown in the box below so that it will know how to compute the temperature with the information being supplied.

Now that we have covered the basics, we will build a few working models, test them, and learn more modeling skills as we work our way through the models in the following sections.

Earth Temperature Equation

Earth_Temperature = Earth_Energy / Earth_Heat_Capacity

HEAT CAPACITY AND THE FIRST LAW OF THERMODYNAMICS

We know from common experience that when we add heat to an object, its temperature increases. We also know that the temperature of air increases when we compress it and decreases when we allow it to expand. This is why the air is cooler in the high mountains, where it is expanded, than at lower altitudes, where it is compressed. The first law of thermodynamics provides a quantitative statement of these observations and restates the law of conservation of energy for systems. The law of conservation of energy states that energy can be transferred between systems and transformed into different forms of energy within systems, but the total energy must remain unchanged. The energy in a system that causes its temperature to increase is called *thermal energy*, E, and the energy that leaves a system when the system expands is called *work-by*, W. (If we computed the energy expended in compressing the system it would be called *work-on*.) The *heat energy added* to the system, Q, must increase the system's thermal energy, dE, plus supply the energy produced by the system in doing work (*i.e.*, energy is conserved).

$$Q = dE + W \qquad (1)$$

We use temperature, T, to measure the thermal energy of a system; thus, the change in temperature, dT, can be computed from a simple equation.

$$dE = C\, dT \qquad (2)$$

In this equation, C is a constant for a particular system, and we see that the thermal energy of a simple system is directly related to the system's temperature. In Equation 1 we see that if W is zero, all of the heat energy added will go into thermal energy. We can force W = 0 by not allowing the system to expand. We keep the volume constant because work involves changing a system's dimensions. The constant C for this situation is called "the heat capacity at constant volume," C_v, and since W = 0 we find

$$dT = dE / C_v = Q / C_v \qquad (3)$$

Only in laboratory situations can we keep the volume of a system constant. In nature the pressure surrounding the system usually remains constant while the system changes volume. The heat capacity for this situation is called "the heat capacity at constant pressure," C_p.

$$C_p = Q / dT = (dE + W)/ dT = C_v + W/dT \qquad (4)$$

In Equation 4 we see that C_p is always greater than C_v because the positive quantity W/dT must be added to C_v to make them equal. For solids and liquids, however, the volume change associated with heating is so small that C_p and C_v are nearly equal. With gases the volume change is larger, so C_p differs from C_v; for air $C_p = (7/5)\, C_v$.

The heat capacity of a system must be proportional to the size of the system; twice as much water must have twice the heat capacity. We therefore define the "specific heat capacity" as simply the heat capacity per unit mass of a given material. A lower case c is used for a specific heat capacity.

Exercises

1. Sketch the basic system-diagram structures for the following systems:
   (a) inventory, sales rate
   (b) power (from any source), energy
   (c) deposits, withdrawals, account balance
   (d) river, lake
   (e) distance an object travels over a fixed period of time, speed

2. Identify the source and sink for a coal-burning power plant.

3. The long-duration spaceship presents a challenge in materials recycling. Using the water supply example as a guide, sketch simplified system diagrams for the air supply and the solid materials (food, refuse, etc.) for the spacecraft. Indicate important connecting flows between the systems. Will the system as a whole require anything external to continue operating? Will the system as a whole release anything to space?

4. Sketch a system diagram for the system "Do I need to buy more milk?" showing the reservoirs, flows, converters, and connectors necessary to make the system run. Explain with words or mathematical statements the contents of all of the components in your system diagram.

5. Three 1-kg objects of different materials (aluminum, carbon, and water) all absorb 1,368 joules of thermal energy. What is the temperature increase of the three objects? The specific heats for the materials are, respectively, 899 J kg^{-1} K^{-1}, 690 J kg^{-1} K^{-1}, 4,218 J kg^{-1} K^{-1}.

IV

Building Working Models: The Bathtub Model

We have employed three working examples to explore the various aspects of modeling. In this section and the two that follow we will complete these examples, run them, and look at the systems' behaviors under different conditions. The three examples are: the bathtub, the Earth energy system, and human population. New terminology and techniques will be introduced as we use them.

In some learning environments STELLA® II software from High Performance Systems, Inc., will be available to the students; in these situations the working models can be built and run following the discussions in the text. STELLA® II for the Apple Macintosh™ and Windows systems can directly read and interpret the model system diagram and equations. If

STELLA® II is not available, some of the material in this section and the ones that follow can be skipped. Such materials are bounded by icons, ↓■ and ■↑. Even if you do not have access to STELLA® II, please read on; you can learn the basics of system behavior from the following sections without it.

The bathtub model on page 15 was nearly complete as a system diagram. We now add three new converters. "Bather D1" tells the model when to turn the fill valve on and when to turn it off; "Bather D2" tells the model when to close and open the drain. (The 1 and 2 refer to decisions, not individuals.) "Bathtub Volume" is a constant that tells the "Overflow" valve the volume of the bathtub below the overflow drain.

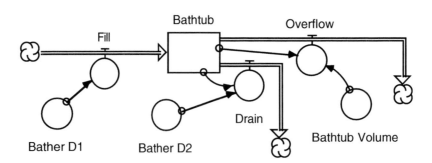

Bathtub Model Equations

1. Bathtub(t) = Bathtub(t - dt) + (Fill - Drain - Overflow) * dt
   INIT Bathtub = 0.0 {m^3}
2. Fill = Bather_D1
3. Drain = IF Bathtub > 0.0 THEN Bather_D2 ELSE 0.0 {m^3/min}
4. Overflow = IF Bathtub > Bathtub_Volume THEN 60e-3 ELSE 0.0 {m^3/min}
5. Bather_D1 = IF TIME < 10.0 THEN 60e-3 ELSE 0.0 {m^3/min}
6. Bather_D2 = IF TIME > 20.0 THEN 120e-3 ELSE 0.0 {m^3/min}
7. Bathtub_Volume = 1*2*0.5 {m^3}

We must provide the model with the diagram on page 20 and tell it how to make the necessary decisions and computations with the equations in the box above.

1. This equation was written automatically by STELLA® II as a statement of the reservoir and flows that we constructed on the system diagram. We had to supply only the initial condition (INIT Bathtub) on the second line; the bathtub starts out at TIME = 0.0 s with no water (*i.e.*, 0.0 m^3 of water). In writing this equation, STELLA® II has recognized that there are three flows involved in the bathtub dynamics. "Fill," which is entering the bathtub, is positive, whereas "Drain" and "Overflow," which are leaving the bathtub, are negative. Equation 1 states that for each time step "dt" we evaluate the flows entering and leaving the reservoir and add that quantity to the quantity of water in the bathtub. If "Fill - Drain - Overflow" is a positive quantity, then "Bathtub" will increase; if it is negative, then "Bathtub" will decrease.

2. The "Fill" valve takes the value provided by "Bather D1." See Equation 5.

3. If there is water in the bathtub, the "Drain" valve will equal "Bather D2"; if there is no water in the bathtub the valve will be 0.0. See Equation 6.

4. This is the overflow logic that was mentioned in our discussion of branches and decisions (page 14). There is zero flow through the "Overflow" valve as long as the bathtub water volume is below the "Bathtub Volume." When the bathtub water volume reaches or exceeds the "Bathtub Volume," the "Overflow" valve is set to 60×10^{-3} m^3 min^{-1}. Here we have included the decision logic in the "Overflow" valve; compare this to the "Fill" valve, where the decision was externalized in the converter "Bather D1." Options like these are up to the individual tastes of the modeler; the best guideline is to make the system diagram as clear as possible.

5. "Bather D1" becomes 60×10^{-3} m^3 min^{-1} from "TIME" = 0.0 until "TIME" = 10.0; after "TIME" = 10.0, "Bather D1" is 0.0. The value of "Bather D1" is equal to the flow rate that the "Fill" valve will have. This decision could have been included internally as a part of "Fill."

6. "Bather D2" is 0.0 before "TIME" = 20.0; after "TIME" = 20.0, "Bather D2" becomes 120×10^{-3} m^3 min^{-1}. The value of "Bather D2" is equal to the flow rate that the "Drain" will have.

7. The assumed dimensions of the bathtub are: width = 1.0 m, length = 2.0 m, and height to the overflow = 0.5 m. These are used here to compute the "Bathtub Volume."

The curly brackets, { }, appearing in the equations contain notes from the modeler and are not considered part of the equation by STELLA® II. I have used them here to document the units of the equations; m^3 for reservoirs and m^3/minute for the flows.

The "if statement" is the most common form of decision making or branching logic used in system modeling. It monitors the continually changing conditions of the model as it runs and can change the action taken by the model based upon current conditions, an important capability in dynamic models. Actually, the "if statement" supplies a value to an equation in the model. Every time the model requests new information the "if statement" reevaluates the situation in the running model based upon the model's current conditions, computes a new value for the "if statement," and gives it to the model. The basic "if statement" has three parts as illustrated below:

IF "test statement" THEN "true value" ELSE "false value"

The "test statement" is a question posed to the model; examples of a "test statement" are:

$x > y$ (Is "x" greater than "y"?), total = 0.0 (Is "total" equal to zero?), level < marker (Is "level" less than "marker"?), $x > a$ AND $x < b$ (Is "x" greater than "a" and simultaneously less than "b"? Or, does the value of "x" lie between "a" and "b"?).

The "true value" and "false value" depend upon whether the answer to the question posed by the "test statement" is true or false; "true value" and "false value" may be constants, variables, statements, or functions. Examples of "true value" and "false value" are: 1,368.0, level, $x - y$, SQRT($x^2 + y^2$). As an example, let the variable "speed" represent the speed of an automobile in the model; the "if statement" below allows the model to adjust the speed of the automobile to approach the legal speed limit.

IF speed < 55.0 THEN +1.0 ELSE -1.0

The model could use the value supplied by the "if statement," either +1.0 or –1.0, to change the speed of the automobile by adding it to "speed," which in this example would add or subtract 1.0 from the model automobile's speed.

Notice that the structure of the "if statement" follows sentence structure; think how the following sentence might appear as an "if statement":

If that animal is a bear we should run; otherwise we can whistle.

In scientific notation, numbers are displayed in a compact standard form by utilizing powers of ten to conveniently position the decimal. Large numbers like 6,371,000.0 m (the Earth's radius) can be written 6.371×10^6 m, and small numbers like 0.00000058 m (the wavelength of yellow light) can be written 5.8×10^{-7} m. This notation works well in written text. Calculators and computers, however, rarely have superscripting capability, so other conventions have emerged. One common convention is to substitute "E" or "e" for the "x10," and then place the "power" immediately following it.

6,371,000.0 m	=	6.371×10^6 m	=	6.371E6 m	=	6.371e6 m
0.00000058 m	=	5.8×10^{-7} m	=	5.8E-7 m	=	5.8e-7 m

You will see these various forms appearing in this module.

When we run the bathtub model we get the results plotted on the graph below.

This graph tells the whole story. The horizontal axis is time in minutes. There are four variables listed across the top of the graph, with a number assigned to each. The vertical axis has four scales with the variable numbers. For example, "Bathtub" (Variable 1) has a scale range from 0.0 to 1.0. The other three variables have scale ranges from 0.0 to 0.12. Each of the four variables is plotted on the same graph, with the variable identifier number printed periodically over the relevant line.

Look first at "Fill" (2). The valve is turned on at 0.0 minutes, and the water flows into the "Bathtub" (1), slowly filling it. At 10 minutes the valve "Fill" (2) is turned off, and the water in the "Bathtub" (1) remains at a constant level for ten minutes. Now look at "Drain" (3). This variable has had the value 0.0 from the beginning until 20 minutes, when the "Drain"

valve is opened and empties the bathtub. Since the "Drain" rate is twice the "Fill" rate, it takes only half as long to empty the bathtub as it did to fill it. What about the "Overflow" (4)? The overflow branch never was activated because the water volume did not reach sufficient size to overflow; thus, "Overflow" has the value zero throughout this run.

This scenario was rather predictable because the conditions were tightly controlled by the bather through "Bather D1" and "Bather D2." When an independent variable controls a model's behavior, it is called a *driver* or a *forcing function*. The term driver is a good one because the driver of a car is in control of the steering, accelerator, and brakes, and has a route or objective in mind. In this bathtub scenario, we might describe the driver as a blind driver because the decisions were based solely on time without any reference to the system behavior.

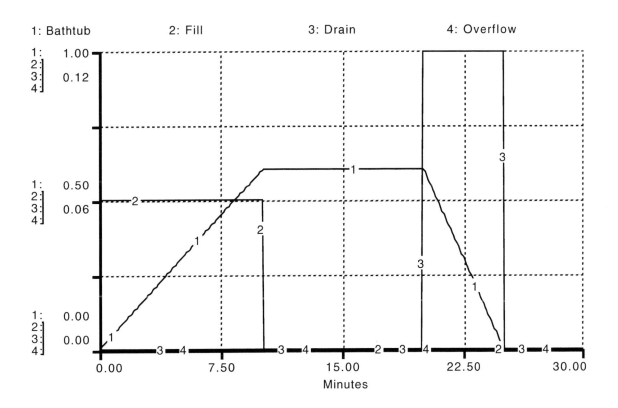

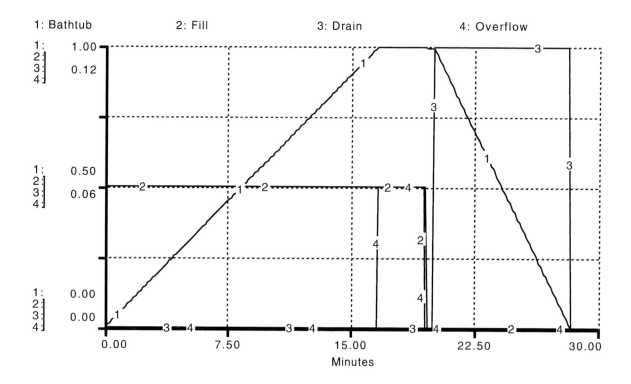

1: Bathtub   2: Fill   3: Drain   4: Overflow

Let's make the bathtub model a little more interesting by changing the turn-off time to 19.5 minutes. (Perhaps the bather received a telephone call and forgot to turn off the water.) The new Equation 5 is below. This minor change in the turn-off time changes the output from the model appreciably.

Revised Equation 5

5. Bather_D1 = IF TIME < 19.5 THEN
                60e-3 ELSE 0.0 {m³/min}

In our previous scenario the "Overflow" was inactive and Variable 4 remained at zero throughout the run; here we see Variable 4, "Overflow," becomes active around 17 minutes and maintains the bathtub at a constant level until 20 minutes, at which time "Bather D2" opens the "Drain" and the bathtub is emptied. When a variable becomes active only after a system parameter reaches some prescribed

value, that value is called a *threshold*; threshold variables are common in Earth systems. When the strain along a fault in the Earth's crust reaches a critical point, for example, it is released in an earthquake. This scenario was more interesting than the previous one because the model took an action not prescribed by the driver.

At precisely what time does the overflow drain become active? We can compute that from the system parameters. Any time we divide a *reservoir* by a *flow*, the quotient is called a *time constant* or *time scale* for the flow. For this example, the "Fill" time constant, T_f, is computed from "Bathtub" and "Fill." Similarly, we may compute the time constants for the "Drain," T_d, and the "Overflow," T_o. In this scenario $T_f = T_o$, so the bathtub level remained constant during the period (16.7 min. to 20 min.) that both flows were active. When time constants increase, the processes involved slow down; when they decrease, the processes proceed faster.

Time Constant Computations

$$T_f = \frac{\text{Bathtub}}{\text{Fill}} = \frac{1.0}{60 \times 10^{-3}} = 16.7 \text{ min.}$$

$$T_d = \frac{\text{Bathtub}}{\text{Drain}} = \frac{1.0}{120 \times 10^{-3}} = 8.3 \text{ min.}$$

$$T_o = \frac{\text{Bathtub}}{\text{Overflow}} = \frac{1.0}{60 \times 10^{-3}} = 16.7 \text{ min.}$$

We described our bather previously as a "blind driver" because the decisions were strictly programmed without information on the system's performance. Now let us provide the bather with information on the progress of filling the bathtub.

In this new system diagram (below) we added a connector to transfer "Bathtub" information to "Bather D1," and we connected "Bathtub Volume" to "Bather D1." When information about a model's output variables is transferred back to, and utilized by, input variables, especially drivers, the connectors are called *feedback* or *feedback loops*. (The feedback connector is labeled in the diagram below.)

 The converter "Bathtub Volume" appears in two places (far left and below "Overflow"). On the left it is drawn as a dotted or grayed circle to indicate that it is a replicated converter, not the original converter with the same name. We could have drawn a long connector from "Bathtub Volume" across the diagram to "Bather D1," but this would be inelegant and in complex models could create a confusing spaghetti-bowl diagram. Instead we replicated "Bathtub Volume" near "Bather D1" and used a short connector. Replicated diagram elements in STELLA® II are called *ghosted* elements because they are drawn as dotted or grayed rather than solid figures. Only the original diagram elements may be edited; the ghosted elements reflect the current value of the original element, but may not be edited directly.

We now need to provide "Bather D1" with a new equation or decision to use the new information. If we assume that the bather wants to fill the bathtub to 75% of its capacity, we use the new Equation 5 in the box below.

Bather D1 with Feedback

Bather_D1 = IF (Bathtub<
            (Bathtub_Volume*0.75)) AND
            (TIME<20) THEN 60e-3
            ELSE 0.0 {m³/min}

Recall that the format of the "if statement" is

IF "test statement" THEN "true value"
ELSE "false value."

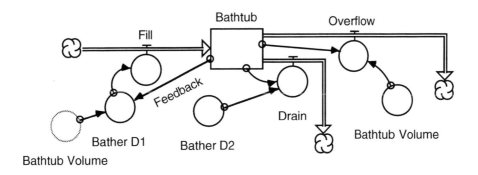

In the "if statement" used above (Bather D1 with Feedback) the "test statement" question has two parts and has the general form "test statement 1" AND "test statement 2"; because of the "AND" both "test statements" must be true for the "if statement" to give the "true value." If the "test statement" had the general form "test statement 1" OR "test statement 2" then only one would need to be true to produce the "true value." ■↑

When we run this model with the feedback loop, we obtain the output displayed on the graph below.

The model is behaving as desired. "Bather D1" turns on the fill valve, monitors the "Bathtub," and turns the valve off when the bathtub level reaches 75% of capacity. In general, when a feedback causes a driver to reduce an output variable, the feedback is called *negative feedback*. (Had the feedback been *positive*, "Bather D1" would have increased the fill valve flow, leading to an overflow.)

Exercise

We have assumed in setting up the bathtub model that "Drain" produced a constant outflow of water when it was opened. Actually, the flow rate should be proportional to the water pressure at the drain orifice.

$$\text{Drain} = (\text{constant}) * \text{Water\_Pressure}$$

The water pressure is proportional to the volume of water in the bathtub.

$$\text{Water\_Pressure} = (\text{constant}) * \text{Bathtub}$$

1. Make a qualitative sketch of how you expect the water volume in the bathtub to behave.
2. Sketch and modify the last bathtub model system diagram to allow for our corrected drain flow rate in "Bather D2." Let the drain flow rate be $(120 \times 10^{-3})(\text{Bathtub}/\text{Bathtub Volume})$.
3. Modify Equation 6 to produce the corrected drain flow rate.

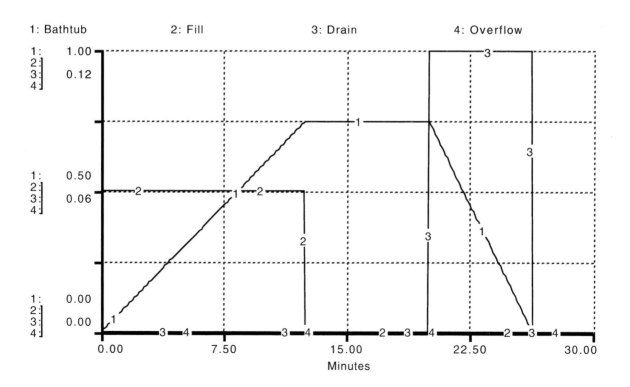

V

Building Working Models: The Earth Energy System

In Section III, we introduced the Earth energy system model and then modified the input side and discussed some changes to the output side. We now add several more elements to produce a working model of the Earth energy system. The physical principle invoked in this model is the conservation of energy. Radiant energy in the form of visible sunlight is absorbed by the Earth's surface; this energy warms the surface and the temperature increases. The Earth radiates energy into space in the infrared region of the electromagnetic spectrum; this loss of energy tends to cool the surface. The model for the Earth energy system conserves the energy flowing to and from the Earth and finds the temperature at which the energy flows are balanced.

We can identify two drivers in this model, "Solar Constant" and "Earth Albedo." Although we are treating both as constants this time, they could become variables.

"Solar Constant," the amount of solar radiation a square meter receives each year at the top of the Earth's atmosphere at the Earth's average distance from the Sun (see Glossary), could change to reflect changes in the Earth's orbit, and "Earth Albedo," the percentage of that radiation Earth reflects back into space, could change in response to global ice cover and global cloudiness. (Both ice and clouds reflect radiation.) Other converters have been added to the diagram to permit the computation of the needed parameters and variables. I have introduced two small separate subsystems

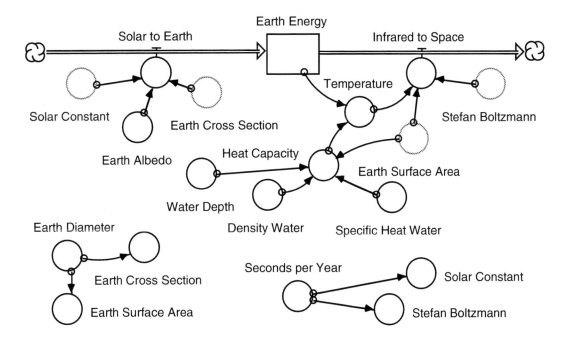

at the bottom of the system diagram. The one on the left uses "Earth Diameter" to compute the "Earth Cross Section" and "Earth Surface Area." These two parameters are ghosted into the main diagram. To the right, the converter "Seconds per Year" is used to convert "Solar Constant" and "Stefan Boltzmann" to annual values. (See equations 10, 11, and 13.)

↓ ▣ All physical models work in time units of seconds, but it becomes awkward to express a year as 31,557,600 s when working with global models. (This number assumes 365.25 days/year to include leap years.) We can use other time units in our models, but we must carefully convert all physical parameters involving time to the new time units.

When a working part of the model is set aside from the main system diagram (the *main program*), we call it a *subroutine*. The output from the subroutine computation can be connected to the main program with a connector, but it is best to ghost it in. Similarly, parameters needed by the subroutine should be ghosted into it.

Below are listed all of the equations and constants used by the model; we will not go through all of them in detail as we did with the bathtub model. All physical parameters are expressed in SI units. (The SI stands for Système Internationale d'Unités, which is the internationally endorsed form of the metric system similar to the MKS [Meter-Kilogram-Second] system.)

The curly brackets, { }, have been used extensively in these equations to document the units involved and other modeler's comments.

We have chosen to store the "Earth Energy" in a one-meter layer of water covering the Earth's surface. This decision is frequently used by global modelers, and such models are called "swamp models" because to simplify the model they treat the Earth's surface as if it had the uniform conditions similar to the surface of a swamp. Equations 4, 9, 12, and 14 are involved

---

Earth Energy Model Equations

1. Earth_Energy(t) = Earth_Energy(t - dt) + (Solar_to_Earth - Infrared_to_Space) * dt
   INIT Earth_Energy = 0.0 {J, We do not know to put here yet. Let the model compute it for us.}
2. Solar_to_Earth = Solar_Constant {J/m2 yr} * (1-Earth_Albedo) * Earth_Cross_ Section {m2}
3. Infrared_to_Space = Earth_Surface_Area {m2} * Stefan_Boltzmann {J/m2 yr K4} * Temperature^4 {K4}
4. Density_Water = 1000. {kg/m3}
5. Earth_Albedo = 0.30 {30% as a fraction}
6. Earth_Cross_Section = PI*Earth_Diameter^2/4 {m2}
7. Earth_Diameter = 12742e3 {m}
8. Earth_Surface_Area = PI*Earth_Diameter^2 {m2}
9. Heat_Capacity = Water_Depth {m} * Earth_Surface_Area {m2} * Density_Water {kg/m3} * Specific_Heat_Water {J/kg K}
10. Seconds_per_Year = 3.15576E7 {s/yr}
11. Solar_Constant = 1368 {J/m2 s} * Seconds_per_Year {s/yr}
12. Specific_Heat_Water = 4218. {J/kg K}
13. Stefan_Boltzmann = 5.67E-8 {J/m2 s K4} * Seconds_per_Year {s/yr}
14. Temperature = Earth_Energy {J} / Heat_Capacity {J/K, 1st Law of Thermodynamics}
15. Water_Depth = 1.0 {m, temporary assumption}

in the computation of the "Temperature." In Equation 9 the mass of the layer of water is computed and multiplied by the specific heat capacity of water to obtain the "Heat Capacity" of our swamp Earth. Equations 2 and 14 were discussed in Section III. Note that in Equation 1 we have set the initial value of the "Earth Energy" at zero; this is arbitrary, but allows us to watch the Earth warm up from absolute zero. ■↑

This model produces the output plotted on the graph below.

The format for this graph is the same as the bathtub graphs. All of the vertical axis scales have been set to place the maximum value of each plotted variable at the top of the graph. This enables us to read the maximum values directly from the upper axis scales.

"Solar to Earth" (3) in the graph is a constant in this scenario; our only reason for plotting it was to obtain its value on the vertical axis scale. "Temperature" (1) and "Earth Energy" (4) are plotted exactly on top of each other. This always happens when two variables are *linearly* related and the plotting scales are normalized

to their maximum values; Equation 14 gives the linear relationship between "Temperature" (1) and "Earth Energy" (4). "Infrared to Space" (2) follows a different curve because "Infrared to Space" is proportional to the fourth power of "Temperature," T^4 (Equation 3). When T is small relative to its maximum value, then T^4 is very, very small, as shown on the lower left corner of the graph; as the two variables approach their maximum values, "Infrared to Space" catches up with "Temperature," and they both slowly merge to their maximum values. This type of system behavior, in which output variables ultimately achieve a constant value and approach that value slowly, is called an *"asymptotic approach to a steady-state solution."*

Looking again at the output graph, we see that the model in the steady-state region predicts a temperature for the Earth of 255 K, or –18° C. This may seem low, but is actually a valid answer, since it represents a global average including the polar regions and the entire atmosphere, where temperature decreases approximately 7° C for every

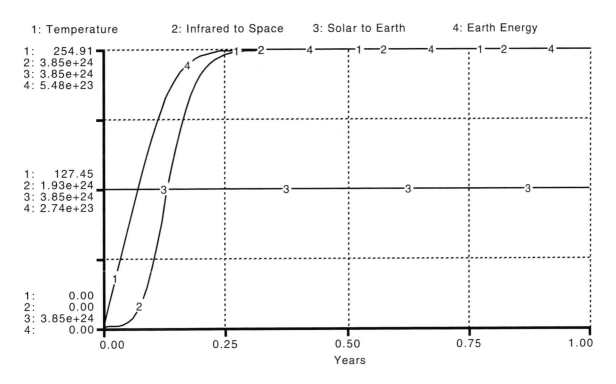

29

kilometer increase in altitude. At the tropopause (the top of the lowest layer of the atmosphere, at about 12 km altitude) the temperature is −60° C! If the Earth's mean temperature were measured from space using infrared detectors, the value would be close to 255 K. The temperature computed using a planet's radiation balance is called the "effective planetary temperature"; it closely matches the planet's temperature measured from space. The

effective planetary temperature is an important parameter for characterizing a planet's relationship to the Sun and its fundamental thermal condition. At this temperature the planet radiates into space exactly the same energy per day that it receives from the Sun. This is a very delicate balance, and any deviation will cause the planet to warm or cool. The effective planetary temperature and the black-body radiation law act together like an overall negative

BLACKBODY RADIATION

A new physical equation is introduced in Equation 3, on page 28, the Stefan-Boltzmann or "blackbody" radiation law. It is:

$$R_{bb} = \sigma T^4$$

The Stefan-Boltzmann constant, σ, is $\sigma = 5.67 \times 10^{-8}$ Wm^{-2}K^{-4}; its value is given in Equation 13 (where the units are also changed to the annual value). The Stefan-Boltzmann law computes the total power radiated per unit area, R_{bb}, from a perfect black material at a uniform temperature, T.

The name "blackbody radiator" seems a strange name to apply to our Earth, which we know from space photographs is predominantly blue, white, and green. The visible colors of the Earth, however, are the reflected light from the Sun, not the radiation produced by the Earth itself. We would need infrared eyes to see the Earth's own radiation, and we would see an entirely different Earth. When we look at an object that absorbs all the radiation that strikes it, it appears completely black. Physicists have proven that all materials radiate electromagnetic energy at each wavelength with exactly the same efficiency that they absorb radiation at the same wavelength (Kirchhoff's law). A "blackbody radiator," which is a perfect absorber, is, therefore, also a perfect radiator of electromagnetic radiation. It is the most efficient radiator possible; it emits the maximum radiation possible at a given temperature, and the distribution of that energy among the various wavelengths of the electromagnetic spectrum follows a specific law that depends upon the temperature.

There are many examples of blackbody radiators in our everyday environment; solar radiation is blackbody radiation. Because of the high temperature of the apparent visible solar surface (approximately 6,000 K), solar radiation occurs primarily in the visible light wavelengths. The incandescent light bulb is another example; the temperature of the filament in the light bulb is approximately 2,800 K, and its light is yellowish white. (When you use a dimmer on an incandescent light bulb you lower the temperature of the filament; the light bulb produces less light, and the light becomes yellow to red as it dims.) White lightning has a temperature about 30,000 K, and is bluish white. The Earth's effective temperature is around 255 K, and its radiation is in the infrared part of the electromagnetic spectrum, at wavelengths much too long to be seen with our eyes.

LINEARITY

The terms *linear* and *linearly related* are frequently used in describing a system's behavior. Basically they mean that one system variable, when plotted on a graph as a function of another system variable, will plot as a *straight line*. This linear relationship has a specific meaning in mathematics. If one variable, *y*, is a function of another variable, *x*, they are linearly related if the algebraic expression describing their relationship can be written: $y = ax + b$, where a and b are constants. The important aspect of this equation is that *x* has the power one (*i.e.*, x^1). Again, the word linear is used because when *y* is plotted as a function of *x*, the result is a straight line. In Section IV, "Bathtub" is a linear function of time while the fill valve is on.

If the algebraic relationship were $y = ax^2 + bx + c$ (c is another constant), we would call it a *quadratic* relationship.

---

feedback process for the planet's climate. Because of the blackbody radiation law (radiated power $\sigma\, T^4$), a small increase in planetary temperature will produce a proportionately much larger increase in outward infrared radiation, which will cool the planet. Similarly a small decrease in planetary temperature will cause a decrease in energy radiated, which will warm the planet.

To obtain a temperature for the Earth's surface that more closely matches the actual surface conditions, we need to modify the model to include an atmosphere, so that *greenhouse warming* of the surface is incorporated. The figure below illustrates the principles involved in the

greenhouse effect. On the left is the situation that we modeled, the Earth without an atmosphere; the incoming solar energy is exactly balanced at 255 K by the outgoing infrared radiation to space.

Suppose the atmosphere lets all solar radiation pass through, totally absorbs all infrared radiation, and has a constant temperature. What is the temperature of such an atmosphere? The answer has to be 255 K, because the Earth ultimately must reradiate to space all of the energy received from the Sun, and our model has computed 255 K as the temperature required to do the job. In a simple model with the atmosphere, it is the atmosphere, not the

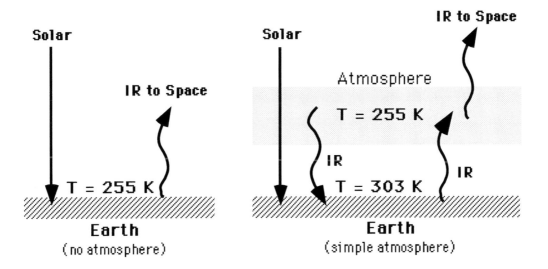

Earth's surface, that radiates into space. Therefore it is the atmosphere that has the 255 K temperature. But the temperature of the surface under this atmosphere is no longer 255 K. Why? Because it is now receiving both solar and atmospheric radiation, and in our simple scenario these two radiation sources are equal. This is depicted on the right side of the figure on page 31. (The atmosphere must radiate as much radiation downward as it does upward, because the molecules are emitting radiation in all directions.) Now the Earth's surface must acquire a temperature sufficient to radiate twice as much energy as before in order to reach a steady state. Our blackbody equation tells us that to achieve twice the radiation power the temperature must increase by the factor $\sqrt[4]{2}$.* Our new value for the surface temperature with an atmosphere is 303 K = 30° C, which is about right for June in Houston, but too high for a global average surface temperature. The reason our simple model gave us a temperature too high was that we assumed a totally opaque atmosphere in the infrared, where it is actually partially transparent. However, by adding greenhouse gases such as carbon dioxide to the atmosphere we decrease the atmospheric transparency in the infrared and make it more opaque.

Let's turn our attention to time constants. We can make a reasonable guess at the warming time constant for the "one-meter swamp Earth" from the graph; approximately three months were required for the Earth to reach the steady-state temperature. But the change is slow during the final two months. If we extend the straight line portion (roughly the first month) of the "Temperature" curve (1), which is also the "Earth_Energy" curve (4), from time = 0.0 until

it crosses the top line representing the steady-state solution, we find a time of approximately 1.5 months. Our guess is that the model Earth warming time constant is between 1.5 and 3 months.

$$T_w = \text{"Earth Energy"} / \text{"Solar to Earth"}$$
$$= 5.48e{+}23 / 3.85e{+}24 = 0.142 \text{ years}$$
$$= 1.7 \text{ months}$$

We compute the warming time constant, T_w, by dividing the reservoir "Earth Energy" (use the steady-state value) by flow "Solar to Earth"; this computation yields 1.7 months, an answer close to that found from extending the straight line portion of the graph. Notice that the cooling time constant, T_c = "Earth Energy" / "Infrared to Space," has exactly the same value as the warming time constant; this is necessary for the system to remain in the steady state.

Our "one-meter swamp" Earth model would be totally useless for studying daily changes in the Earth system because the water averages out all thermal changes occurring in periods of less than a month. But it could be modified to explore seasonal or longer changes in the Earth system because the water could respond to changes occurring over periods longer than 1.7 months.

Why did we use one meter of water? It was an arbitrary choice. One of the really satisfying things about creating a working model is that it can easily be modified to try other ideas. We can, for instance, change the depth of the water or change the water to rock and see what happens. If we adapt our "swamp Earth" to create two new Earth models, one with a half-meter layer of water and the other with five meters of rock, and run all three models, they all reach the same final temperature: 255 K.

---

*Let R_1 and R_2 be the blackbody radiant powers emitted at temperatures T_1 and T_2; then we have $R_1 = \sigma\, T_1{}^4$ and $R_2 = \sigma\, T_2{}^4$ when we apply the blackbody radiation law. If we now divide the second equation by the first we get $R_2 / R_1 = (\sigma\, T_2{}^4)/(\sigma\, T_1{}^4)$, which simplifies to give us $R_2 / R_1 = (T_2 / T_1)^4$ or $T_2/T_1 = \sqrt[4]{R_2/R_1}$. For the simple greenhouse model considered here, R_2 is $2R_1$; hence $T_2 / T_1 = \sqrt[4]{2}$.

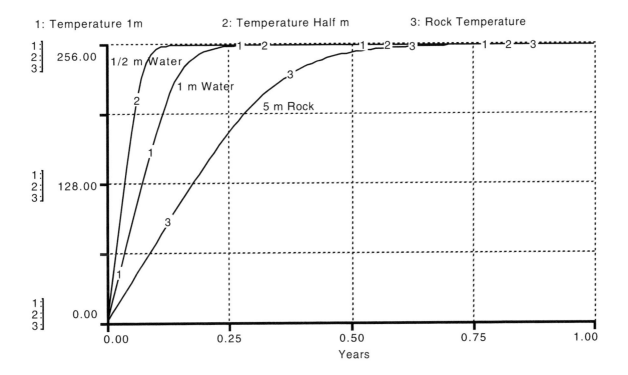

1: Temperature 1m   2: Temperature Half m   3: Rock Temperature

This is not surprising, because it is a steady-state model and we have not changed the two drivers, "Solar Constant" and "Albedo." The difference is in how long it takes the models to reach this temperature. The time constant for a half-meter of water is exactly half that for one meter of water; there is half as much water to heat up, so it can reach any particular temperature twice as fast. The time constant for five meters of rock is approximately four months. If we had used an Earth-sized rock in the model, we would still have ended up with a temperature of 255 K, but it would have taken a long time to reach a steady-state solution (which is one reason for using a meter of water). Another interesting comparison is to find the thicknesses of various substances that have the same warming time constant; the values in the box that follows will probably surprise you.

Now let's ask the question in a different way. Suppose we wanted a model with a surface that would respond to the daily heating of the Sun. How thick should we make the surface? We

> **Thicknesses for Equivalent Warming Time Constants**
>
> 1 meter of water = 2 meters of rock = 4,500 meters of air

know that a meter of water has a time constant of 0.142 years; therefore, a centimeter of water will have a time constant of 0.00142 years, or 12.4 hours, still a little too long to closely follow daily solar heating. The daily thermal variations between night and day do not penetrate more than a few centimeters into the surface of this planet. And seasonal variations, summer to winter, all occur in the upper few meters.

In the oceans, currents and waves keep the upper hundred meters of the water mixed; the ocean temperature does not change significantly with the seasons. The atmosphere, however, responds to changes on all of these time scales. One centimeter of water has the same time constant as 45 meters of air, and seasonal variations are felt through the entire troposphere, 12 kilometers thick.

We saw in the swamp Earth model that the time constants for heating and cooling were equal in the steady-state solution. Now imagine that we have a switch to turn the Sun on and off. Our model output tells us that when we turn the Sun on, the Earth's temperature increases with a time constant of 1.7 months. When the Sun is turned off, the Earth's temperature must decrease at the same rate. The Earth has stored the solar energy, and cannot release it faster than is allowed by the time constant. In the swamp model there is always a delay of 1.7 months between a solar input change and the readjustment of temperature to the new steady state. This delay, or shift in the output variable (temperature) relative to the driver (solar input) is called a *phase shift*. We know from experience that the hottest month of the year is usually August in the northern latitudes but that the largest daily input of solar energy occurs at solstice on June 22, when the Northern Hemisphere is at its maximum tilt toward the Sun. Similarly, the hottest part of a summer day may occur several hours after noon. These phase shifts occur in many variables in many systems, and can usually be traced to some time constant related to a reservoir and a flow.

The time constant of a system plays an important role in how the system responds to system drivers with different time periods. Consider the situation in which the system driver is changing much faster than the time constant of the system. The system in this case responds by reducing or damping its response to the driver; the system tends to average the effect that the driver tries to produce. A meter

GLOBAL WARMING CONTROVERSY

In our model and discussions we have included only the science that relates to radiation laws, and we found a direct relationship between the energy received by the Earth's surface and the resulting steady-state temperature. For the real Earth, processes like ocean currents or volcanic eruptions influence the measured global average temperature. Over the longer time scales (longer than decades) radiation processes will have the dominant influence.

At least part of the global warming controversy is semantics. Warming to the scientist means increasing the radiant energy to the Earth's surface; warming to many others means increasing the temperature. Consider a pot of water on the stove. As heat is added the temperature rises, but when the water starts boiling the temperature remains constant. Would you say that you were no longer warming the pot? The scientists would say that the warming is continuing as before because heat is still being added to the pot; others might respond that you are no longer warming the pot because the temperature is no longer rising. Scientists look at the Earth and say with certainty that increasing greenhouse gases cause global warming because they know that additional greenhouse gases increase the infrared radiant energy from the atmosphere to the Earth's surface. In fact, there is as much scientific certainty in this conclusion as there is in the law of gravity. Scientists are less concerned about the year-to-year changes in global average temperature because these variations are normal and expected. The use of global annual average temperatures as proof of global warming is fraught with problems because they are difficult to measure and show a lot of natural variability and because it takes a long time to demonstrate an unambiguous temperature increase using rigorous statistical techniques.

thickness of water will exhibit a small temperature change in response to daily cycles in solar radiation; yet the water will achieve an average temperature corresponding to the average solar input over a season. If the system driver changes much more slowly than the system time constant, the system will follow the changes in the driver in a continuously evolving steady state. The average temperature of a one-meter layer of water will gradually change with the seasonal changes in solar radiation.

In systems capable of natural oscillations, a special response can occur when the driver period matches the system time constant. In this case the system's response is greater than it would be to the same driver operating at slower periods. This response is often called a *resonance response*. We can illustrate all three responses with a glass half filled with water. The glass of water is the system, your hand the driver. To determine the system time constant, push the glass quickly to one side. The water sloshes back and forth with a certain period, which is the system time constant. When we move the glass across the table more slowly than the time constant, the water follows along with little sloshing. If we wiggle the glass rapidly back and forth at periods faster than the time constant, we can create lots of small waves in the glass, but the average height of the water in the glass is unchanged. (You need to move your hand fairly fast to make sure that you are faster than the sloshing time constant.) Finally, when we move the glass back and forth at a period close to the system time constant, we observe the sloshing amplitude grow and eventually the water sloshes out of the glass.

In a large complex system like the Earth, there are many subsystems and components with many different time constants. The output from such a system has a lot of natural variability; for a given set of drivers the system will approach a steady state, but superposed on it will be natural *fluctuations*. The system components with time constants matching driver periodicities will exhibit the largest regular responses; the other system components produce apparently random variations sometimes referred to as *noise*. It is only the average that is steady in the steady state of a complex system. One needs to look no further than the weather to find a perfect example of large fluctuations superposed upon a steady state. In fact, the weather fluctuations are so large that defining their averages becomes very difficult.

Exercises

1. The steady-state solution to the Earth energy problem corresponds to the condition in which the incoming solar radiation is exactly balanced by (equal to) the outgoing infrared radiation. We can write this steady-state solution as an algebraic expression using previously defined parameters.

$$S (1-A) \pi r^2 = \sigma T^4 4\pi r^2$$

This equation simplifies to

$$S (1-A) = 4 \sigma T^4$$

The temperature in this equation is called the effective planetary temperature; for the Earth this is T_E, where $T_E = 255$ K. We now form a difference equation from the steady-state solution by allowing S and A to be variables. (You may think of the difference equation as the time derivative of the equation multiplied by the time difference, dt.)

$$dS (1-A) - S\, dA = 4 \sigma 4T^3 dT$$

We divide this equation by the previous equation (left-hand side by left-hand side and right-hand side by right-hand side), and rearrange the resulting equation.

$$\frac{dT}{T} = \frac{1}{4} \frac{dS}{S} - \frac{1}{4} \frac{A}{(1-A)} \frac{dA}{A}$$

This form of the difference equation expresses the fractional change in the Earth's steady-state temperature as a function of the fractional change in the solar constant and the fractional change in the Earth's albedo. Equations written in this form are useful for sensitivity studies. Note that the solar constant variation term is positive, corresponding to an increase in temperature associated with an increase in solar constant. The Earth albedo variation term is negative because increases in albedo produce decreases in the Earth's temperature.

The sensitivity of the Earth's temperature to changes in the solar constant is measured by the increase in temperature produced by a 1% increase in the solar constant.

a. Find the sensitivity of the Earth's temperature to solar constant changes.
b. Find the sensitivity of the Earth's temperature to changes in the Earth's albedo (again, for a 1% change in albedo). Assume a steady-state albedo, A = 0.3 (30%).
c. Find the temperature sensitivities of Venus (A= 0.71 [71%]) and Mars (A= 0.17 [17%]) to albedo changes.

2. We can modify the first equation in Exercise 1 to include greenhouse warming of the surface at temperature T_S (=288 K) by an atmosphere at temperature T_A. See the figure for the greenhouse model on page 31.

$$S (1-A)\, \pi\, r^2 + a\, \sigma\, T_A{}^4\, 4\, \pi\, r^2 = \sigma\, T_S{}^4\, 4\pi\, r^2$$

This equation expresses the energy balance at the surface, which must occur for the steady-state solution. The new second term on the left-hand side of the equation represents the infrared radiation incident on the Earth's surface from the atmosphere. We have introduced a new parameter "a," which is the effective gray-body absorptivity for the atmosphere. (A gray body is similar to a blackbody but less efficient by the factor "a.") We have also used the property (known as Kirchhoff's law) that a material emits radiation at a given wavelength with the same efficiency with which it absorbs radiation at that same wavelength. The factor

"a" can take values between 0.0 and 1.0; a = 1.0 corresponds to a blackbody, and a = 0.0 would be an atmosphere totally transparent to infrared radiation and also incapable of emitting infrared radiation. The factor "a" is directly related to the amount of greenhouse gases in the atmosphere.

The first equation in Exercise 1 is the equation that defines the effective planetary temperature T_E (=255 K); so, we may use this definition to replace the first term in the equation above with $\sigma\, T_E{}^4 4\pi\, r^2$. Our equation can now be expressed in a rather simple form.

$$T_E{}^4 + a T_A{}^4 = T_S{}^4$$

This is still the energy balance equation for the surface, but we are using temperature variables to simplify the form of the equation. We now want to write a similar equation for the energy balance that must occur in the atmosphere.

$$a\, T_S{}^4 = 2\, a\, T_A{}^4$$

The left-hand side of this equation is the radiation energy (per square meter) from the surface (at T_S) that is absorbed in the gray-body atmosphere (with efficiency a). The right-hand side is the total radiated energy (per square meter) emitted by the atmosphere; the factor 2 appears here because the atmosphere radiates equal amounts of energy upward into space and downward to the surface. You may want to think of the atmosphere as having two surfaces of equal area, one facing upward and one facing downward. Several algebraic steps were left out in developing this last equation; you should fill in the missing steps.

We can use this last equation to eliminate the atmospheric temperature T_A from our previous result. Fill in the missing steps.

$$T_E{}^4 = \left(1 - \frac{a}{2}\right) T_S{}^4$$

Following the methods that were used in Exercise 1, we now form a difference equation allowing T_S and "a" to be variables but keeping T_E constant. When the resulting equation is written in the fractional format for sensitivity analysis, we have the following result.

$$\frac{dT_S}{T_S} = \frac{1}{8} \frac{a}{(1 - \frac{a}{2})} \frac{da}{a}$$

Specifying "a" for our model atmosphere is somewhat difficult, because the real atmosphere has many layers at many different temperatures, rather than the single one used here, and the Earth's atmosphere has clouds that come and go at several different levels. When an overall average for the outgoing radiation for the whole Earth is determined, we find that 7% of the outgoing radiation comes from the surface, with the balance of 93% from the atmosphere, including the clouds. To use this information we need to write an equation for the fraction of the outgoing radiation that originates within the atmosphere.

$$\frac{aT_A^4}{aT_A^4 + (1-a) T_S^4} = 0.93$$

The numerator, you will recognize, is the outward radiation (per square meter) from the atmosphere. The denominator is the total outward radiation; the second term is the surface radiant energy that did not get absorbed in the atmosphere. With the help of the third equation in this problem, solve the above equation for "a." (Hint: Form the ratio T_A^4 / T_S^4 in both equations.)

Using the value that you found for "a", find the sensitivity of the Earth's surface temperature to changes in the effective gray-body absorptivity (again use a 1% change in "a").

Compare your result for the sensitivity of the Earth's surface temperature to changes in the effective gray-body absorptivity to your results from Exercise 1 for the sensitivities to solar constant and albedo changes. Considering that "a" is the result of atmospheric greenhouse gases, comment upon the relationship of humankind's alteration of the global concentration of greenhouse gases to the natural global energy balance.

VI
Building Working Models: Human Population Model

In the model of human population introduced in Section III, the annual births and deaths are not constants. We have, therefore, added to the system diagram two new converters, "Births per 1000" and "Deaths per 1000," the birth rate and death rate for each thousand people in the population. Demographers refer to these as the crude birth rate and the crude death rate. These are not constants either, over long periods of time; however, their changes are small compared to other parameters in our simple population model. This working model demonstrates many of the important dynamic features of constant growth-rate systems, in which the fractional or percentage increase in size is constant in every equal time interval.

We have also connected "Human Population" to the birth and death valves because we will need this number when computing "Annual Births" and "Annual Deaths."

An almanac is a good source for actual numbers to put into this model. But first, let us see if we can estimate them. The procedure we will employ is fondly called "the educated guess" or "guesstimate" by scientists. Guesstimating is frequently used by model builders because the numbers that are needed in the model are not always available, and the educated guess will allow the model to run until better numbers can be produced.

In a group of a thousand people, some will be younger than the normal childbearing age and some older. If we assume that the three groups are roughly equal in number, we will have 333 persons of childbearing age. Births occur to couples in this childbearing subpopulation—roughly half of 333, or 167 couples. If we now assume that each couple has two children during the childbearing years, then 333 births will occur over that period. In round figures, we can take the childbearing period to be the 20-year period from 20 to 40 years of age. Dividing 333 births by 20 years yields 17 births per year per 1,000 persons; this is our "educated guess" for the crude birth rate.

In order to "guesstimate" the crude death rate, we assume that all deaths occur in the older population group of 333 persons and that these deaths are spread over the 40-year period from 40 to 80 years. This yields a crude death rate of 8 deaths per year per 1,000 persons. When we run the model, however, it will yield results that are inconsistent with the assumptions used in our guesstimation; we assumed

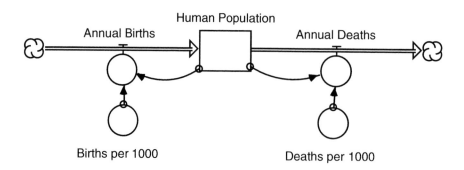

Human Population

Annual Births

Annual Deaths

Births per 1000

Deaths per 1000

a static population distribution and the model produces a dynamic solution.

The final values that we need for the model are the initial value for the "Human Population" and a starting time for the model. We will take these to be 5 x10^9 (5 billion) and 1990. The equations for the population model are given in the box below.

Let's run the model for an 800-year period starting in 1990.

Population Model Equations

1. Human_Population (t) =
   Human_Population (t - dt) +
   (Annual_Births - Annual_Deaths) * dt
   INIT Human_Population = 5e9
2. Annual_Births = (Human_Population/ 1000)*Births_per_1000
3. Annual_Deaths = (Human_Population/ 1000)*Deaths_per_1000
4. Births_per_1000 = 17
5. Deaths_per_1000 = 8

In approximately 750 years the human population will increase to 4 x10^{12}—1,000 times today's population. Imagine Boulder, Colorado, a city of 80,000, becoming a city of 80,000,000; Houston, Texas, growing to 2,000,000,000 (that is the equivalent of half of today's total Earth population moving to Houston). If 4 x10^{12} persons were spaced evenly over the land surface of the Earth, there would be a person every 6 meters (20 feet), ocean to ocean, pole to pole, on rivers, deserts, mountains, highways—everywhere on land—without a gap anywhere. That's too close for comfort!

These numbers seem much too large; perhaps we erred in making our educated guess. The actual figures for the world population in 1990 are: crude birth rate = 27, crude death rate = 10, population = 5,321,000,000. We erred, but on the low-growth side, not the high side. The assumption of two children per couple is typical of most industrialized nations, but too small for the developing countries and the world as a whole. And an 80-year life span is too long for the world population. The rates we guessed are close to the actual figures for the United States

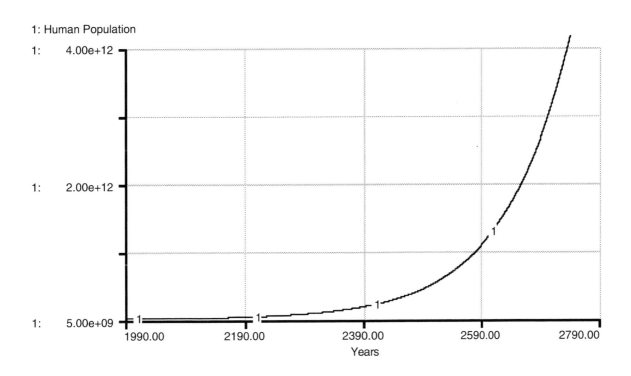

1: Human Population

(in 1990 the crude birth rate was 16 and the crude death rate was 9); this demonstrates the cultural influence on the numbers we selected.

Let us replace our guesstimates with the actual figures and rerun the model. The graph below shows the new results plotted as an overlay on our previous result.

Now that's frightening. Using actual population growth figures for today, we reach the 1,000-to-1 growth factor in 350 years, not the 750 years our guesstimate yielded. If we allow the new model to run for 750 years, we obtain a growth factor of 1,000,000-to-1, a million new people for every one alive today. If spaced evenly over the land surface of the Earth, we would stand 20 cm (8 in.) apart.

Clearly the Earth cannot sustain human growth at 1,000 to 1, much less 1,000,000 to 1. Resources, food production, pollution, and other human-impacted components in the Earth system will combine to enforce *limits* and *constraints* on population growth that we have not included in our population model. In the

bathtub model, the overflow drain represented an explicit constraint on the water level. There was also an implicit limit: the top of the bathtub. We have implicit constraints for the human population model, but they are not as clearly defined.

In our simple population model, we have only two parameters that we can alter to constrain population growth: crude birth rate and crude death rate. If we can act as an intelligent global community, we can control the crude birth rate. We have done so in some industrialized countries. If we do not behave as an intelligent global community, then nature will act to constrain population growth. Nature can play tough when faced with tough problems; the classic options for population control are war, plague, and famine. When nature is the dealer in the population game, it's the crude death rate that is the suit of choice. (More information on population dynamics is included in the module *Population Growth*, in this series.) When we examine the graphs on pages 39 and

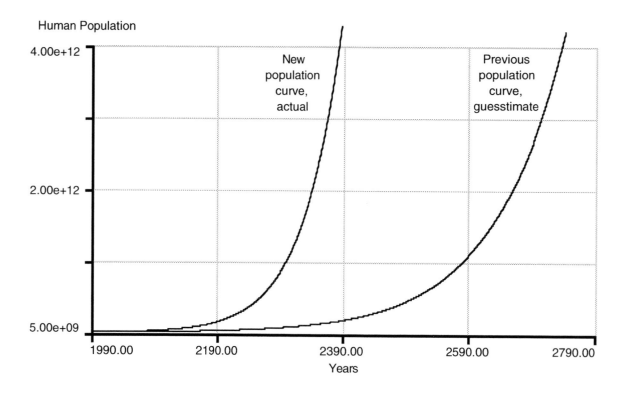

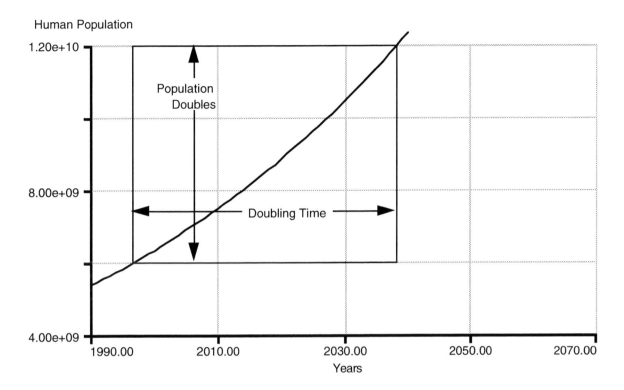

Human Population

40, we might be misled into believing that there is very little change during the first 100 to 200 years because of the graphic scale. We can get a close-up view of the next 80 years by expanding that portion of the graph.

The graph above brings the population problem close to home. The population curve goes off the top of the graph at 12 billion humans, within your expected lifetime. That is approximately three persons for every one in 1980, when the population stood at 4 billion. It means tripling the numbers of cities and further encroaching on the agricultural land, or tripling the living density within our present cities. Imagine sharing your living space with two additional persons, encountering three times as many cars on your city streets, and attending classes three times as large.

What is it about population that makes it behave so differently from the bathtub and the Earth energy system models, which reached a maximum size and then stopped growing? The modelers will respond, "positive feedback"; the demographers will respond, "constant growth rate"; and the mathematically oriented will respond, "exponential growth." These terms are all correct, and they all address the cause of the growth, but with slightly different emphasis.

When the modeler says feedback, she or he is referring to the connector from "Human Population" to "Annual Births" in the system diagram; the affected components are reproduced below. What determines whether the feedback is positive or negative is the equation computing "Annual Births," Equation 2.

The term in this equation containing "Human Population" has a positive sign; therefore, the feedback is positive. An increase in population causes an increase in births, which causes a further increase in population,

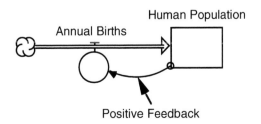

which . . . A positive feedback model increases or decreases until the system reaches some system limit. Positive feedback models sometimes appear to race between limits. During recent geologic time, the Earth's climate has exhibited characteristics of a positive feedback system in its repeated transitions from ice ages to interglacial periods and back.

The demographer would add that the growth rate (annual percent increase in population) is constant. For the population model the "annual percent increase in population" is found by subtracting "Annual Deaths" from "Annual Births," dividing the difference by "Human Population," and then multiplying by 100 to get percent. Equations 2 and 3 yield

---

EXPONENTIAL FUNCTIONS

When the rate of change of x can be expressed

$$\frac{dx}{dt} = ax, \ a = \text{constant}$$

then x has the solution

$$x = x_0 \, e^{at}$$

where x_0 is the initial value of x (x at t = 0). This function is the exponential function, and it is always the solution to the situation in which the change in a variable is directly proportional to size of the variable. The "e" that appears in the exponential function is a constant, e = 2.718281828459.

e-folding Time

The "e-folding time," T_e, is the time required for x to increase by the factor "e"; this occurs when $aT_e = 1$ because $e^1 = e$. Solving for the e-folding time, T_e, we get

$$T_e = 1/a$$

Doubling Time

As you might guess, the doubling time, T_d, is the time required for x to double in value; the exponential function tells us that for this to occur,

$$2 = e^{aT_d}$$

We solve this expression to obtain the doubling time, T_d.

$$T_d \ = \ \frac{\ln 2}{a} \ = \ \frac{0.693}{a} \ = \ 0.693 \, T$$

The doubling time is just a constant fraction of the e-folding time (~70%). If we choose to represent "a" as a percent rather than a fraction, a%, then our expression for the doubling time may be approximated as

$$T_d \ = \ \frac{70}{a\%}$$

In our population model a% is the growth rate (1.7%); hence, the population doubling time is 41 years.

("Births per 1000"–"Deaths per 1000")/10 or, if you prefer, (crude birth rate–crude death rate)/10. If we use the actual population figures in our model we get a 1.7% growth rate. Amazingly, a measly 1.7% growth rate will produce a tripling of the 1980 global population by the year 2040.

The mathematician points out that this is all a consequence of the exponential function. Any time the rate of change of some quantity is proportional to the quantity itself, the function for the quantity is an exponential function. In our model, the difference ("Annual Births"–"Annual Deaths") is the rate of change of "Human Population," and we see in Equations 2 and 3 that the rate of change is proportional to "Human Population." "Human Population" is, therefore, described by an exponential function. The exponential function looks like our population graphs; it always seems to be going off the top of the graph.

An important point to make before leaving this subject is that a constant growth rate produces an exponential solution, a relatively simple mathematical function. However, if the growth rate is not constant but remains positive, we will still get a rapidly increasing growth in population, but the function will not be a simple exponential. The only way to get the population curve to stop growing is to have a zero growth rate, which means the birth and death rates must be equal. The only way to decrease the population is to achieve a negative growth rate, which requires the death rate to exceed the birth rate. There are not many volunteers for this scenario.

Since our model is simple, this analysis is also oversimplified. Consider the additional fact that there is an approximate 20-year phase lag between birth and childbearing age, and another 20 years are required to repopulate the childbearing age group. If we were to achieve zero growth rate today and keep it at zero, it would require 40 years for the population curve to flatten out. In that period the population can double in size. The world has not achieved zero growth rate, although some progress is being made in this direction.

Having brought up the subject of time, let's return to our earlier discussion of model time constants. Following the procedures described earlier, we can compute the time constants for "Annual Births" or "Annual Deaths" by dividing the reservoir "Human Population" by the respective flows. It is more important, however, to know the time constant for population growth, which we can obtain from the difference between "Annual Births" and "Annual Deaths." The equation for the time constant for growth, T_g, is given in the box below.

In our model this works out as 59 years [1000/(27-10)]. Demographers, however, prefer to use a different measure of "time to change," called the doubling time: the time required for the population to double. The doubling time is illustrated on our last population graph, in which we have indicated the region where the population doubles from 6 billion to 12 billion. We estimate from the graph that this occurs over a period of 41 years. (For exponential functions the doubling time is approximately 70% of the time constant for growth.) The doubling time is a very useful concept because it tells us how much time we have to provide for a doubling of the population. Our model results say that we have 41 years to double all of the world's resources utilized by humans, just to maintain the status quo.

Equation for Population Growth Time Constant

$$T_g = \frac{\text{``Human Population''}}{(\text{``Annual Births''}-\text{``Annual Deaths''})} = \frac{1000}{(\text{``Births per 1000''}-\text{``Deaths per 1000''})}$$

Exercises

1. In our guesstimate scenario we assumed the population was divided into three equal-sized groups, aged 0-20 years, 20-40 years, and 40-80 years, and that each couple produced only two children. This "birthing strategy" should have produced zero population growth (ZPG) and a steady-state population, but it did not. Why?

   Our assumptions gave us an estimated crude birth rate (CBR) of 17 and an estimated crude death rate (CDR) of 8; we cannot achieve ZPG until these two numbers are equal. We need to decrease the annual births or increase the annual deaths, or both, to achieve ZPG. Is there a solution to this problem that is consistent with the two-children-per-couple birthing strategy? Explain.

   In a steady-state population there is ZPG and an unchanging age distribution. If there are no significant deaths prior to age "D," the fraction of the population in each age group of equal period prior to age "D" must be the same. Explain why this must be true.

   Let "f" represent the fraction of the population in the childbearing age range, assumed to be 20-40 years. From your arguments in the previous paragraph, "f" also equals the fraction in the 0-20 age range of a steady-state population. What is the fraction in the over-40 group? Using the guesstimating technique employed in the text, estimate the CBR and the CDR as functions of "f." For what value of "f" will the CBR equal the CDR?

   The guesstimating techniques employed in this exercise are simply educated guesses. We have seen that, although useful, they produce results in dynamic models that are inconsistent with the original assumptions. The value that you found for "f" in the previous paragraph may not exactly equal the value that one would obtain by running a sophisticated dynamic computer model, but

   your result illustrates essential features of modern ZPG and near-ZPG populations: the fraction in the child-bearing group shrinks, and the fraction in the older group grows. One inevitable consequence of ZPG societies in the next century will be that the fraction of dependents (under 20 and retired) will grow to approximately one-half of the population. This demographic change has wide implications to social systems; consider voting power and social security taxes as examples.

2. We can present the population age distribution in a graphical format in which the age of individuals is plotted on the vertical axis and the percent of individuals in each age group is plotted on the horizontal axis with males on the left and females on the right. Demographers call this a population pyramid. The population pyramid (top, next page) reflects our assumed population distribution in the guesstimate scenario. The areas of the age groups are all equal, reflecting our assumption that each group contained the same number of individuals. The sizes of each group for each 1,000 in population are listed in the column to the right in the figure.

   A cohort (the group of all individuals born in the same year) starts life on the bottom row on the pyramid and each year moves up one year. Each year a newborn cohort is added to the bottom row of the pyramid. As individuals die, the number in their cohort decreases and the sides of the pyramid shrink toward the center axis, as demonstrated in the section of the population pyramid representing ages between 40 and 80. A steady-state population is one in which there is ZPG and an unchanging age distribution; for a steady-state population the size and shape of the population pyramid are constant in time. Convince yourself that, when deaths from the 0-40 age group are ignored, the under-40 part of the population pyramid must be rectangular for ZPG and that in the cohort-declining years (beyond 40 years in the above example) the number per

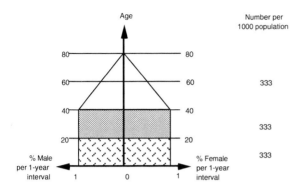

cohort must either decrease or stay constant with age.

The problem with the population pyramid depicted above is that half the persons in each cohort must die by age 60. This does not reflect the real situation, and would be unacceptable in modern societies. The population pyramid below more closely

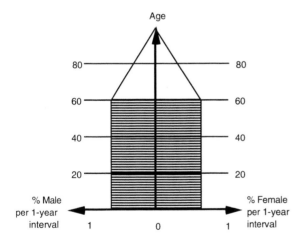

exhibits the population age distribution for an ideal steady-state (and ZPG) modern (or future) society.

In this population pyramid half of the persons in a cohort die before age 80 and the remainder die by age 100. This is similar to conditions in some societies today (the life expectancy at birth in Iceland is 80 years; in the U.S. and Sweden it is 79 years; and for the whole world it is 67 years, based on 1992 population data). Using this idealized steady-state population pyramid find the following quantities.

1. For 1,000 persons in this population, how many are aged 0–20 years?

2. For 1,000 persons in this population, how many are aged 20–40 years?

3. For 1,000 persons in this population, how many are aged 40–60 years?

4. For 1,000 persons in this population, how many are aged 60–80 years?

5. For 1,000 persons in this population, how many are aged 80–100 years?

6. What is the crude birth rate (= crude death rate) for this population?

7. What fraction of the population is "dependent" (under 20 or over 60)?

8. What fraction is at risk with respect to high medical expenses?

VII
Model Characteristics

In the previous three sections, we built and ran three different models, observed their behaviors, and modified the model parameters to change the behavior. Our three models exhibited radically different behaviors. The bathtub model was a linear model and was highly predictable because it was a driver-dominated model. There was only one internal model parameter, "Bathtub Volume," that was a constant, and one variable, "Bathtub." All of the model variability originated with the driver, "Bather," and the one internal logic element, "Overflow." The Earth energy system model was a steady-state model, and it too had a predictable outcome, a simple algebraic solution. The important characteristics of this model are that it is a reasonable representation of a real global physical process and that by changing the model parameters we were able to observe the response of the system. The Earth energy system model served as a tool to help us understand how the system works. Had we altered other parameters, such as "Earth Albedo" and "Solar Constant," we could have used this model to perform sensitivity studies of the response of the Earth's temperature to these changes. The human population model was an exponential model; it also had a mathematical solution. Whereas the two previous models were driver-dominated, the human population model was completely dominated by internal system dynamics. The only driver was crude growth rate ("Births per 1000" – "Deaths per 1000"), which we treated as a constant. We found the time constants changed as we altered the crude growth rate, but the shape of the population

curve was always, unavoidably, exponential growth.

Feedback is an important system concept, and all three models illustrate the behavior of feedback in systems. The human population model had positive feedback, with the result that the population grew without apparent bounds. Although we did not include the system constraints on the population, we recognized that eventually something in the real system would act to limit the growth. This is the characteristic behavior of positive feedback systems; they evolve to some limiting state, and some systems will oscillate between limiting states. Negative feedback, on the other hand, produces a stable steady-state result. In one of the bathtub model scenarios, negative feedback allowed the "Bather" to achieve and maintain a bathtub level of precisely 75% of bathtub capacity. The Earth energy system model did not employ true feedback; however, the "Infrared to Space" valve behaved like a negative feedback component. When the "Earth Energy" increased, the "Infrared to Space" valve caused an increase in the rate of release of that energy. The Earth energy system model, therefore, achieved a steady-state solution similar to negative feedback models.

Many complex models will have several feedback loops, some positive and some negative; in these models the various feedbacks compete. As the system slowly evolves through a series of states, one or another feedback may temporarily become dominant. Some climatologists think that the climate history of the Earth is an example of this type of system behavior. During the Pleistocene, about 10,000 to 1.6 million years ago, a positive feedback process may have been influencing climatic changes, causing the great ice sheets to advance and retreat. There was also an active process in the system that limited the extent of the ice sheets. The Cretaceous, a long warm period in Earth history (65 million to 145 million years ago), may have been under the influence of the positive feedback of the greenhouse gases water vapor and carbon dioxide. Over the very long period of Earth history the planet has maintained a relatively constant temperature; it has not frozen over, nor have the oceans evaporated. For this long period, we can say with certainty that negative feedback processes have dominated because only a system dominated by negative feedback can achieve a steady state intermediate between possible extremes. One of those processes is the infrared radiation emitted to space, which has an overall control over the effective planetary temperature on both long and short time scales (see the discussion of effective planetary temperatures in Section V). Another long-term process has been the negative feedback provided by planetary CO_2 and oxygen chemistry. Very early in the Earth's history the solar output was ~20% lower than today. If the Earth's atmosphere had not been richer in carbon dioxide than it is now, the Earth would surely have frozen over completely and remained so until today.

All three models were simple models; they each had a single reservoir, and they all had analytical mathematical solutions. (An analytical solution is one that can be expressed with an algebraic equation.) Even though the analytical solutions were available, we found that the modeling approach provided additional insight into the workings and behaviors of these systems. The more important point is

that if we change some of the model parameters that we had treated as constants, such as "Earth Albedo" or "Births per 1000," into the variables they are in the real world, the analytical solutions no longer work. The models, however, keep on working. One powerful argument for computer modeling is that the real-world properties need not be overly simplified in order to constrain the problem to fit a simple mathematical format.

When we start adding a higher level of complexity to our models, such as atmospheric layers, carbon dioxide, and clouds to the Earth energy system model, or famine, education, contraception, and resources to the human population model, the analytical approach completely fails us. Modeling is the only method to quantitatively solve these complex real-world problems.

There are always detractors who will discount the results of models. Their objections usually fall into two categories: "But, you didn't include such-and-such a process in your model," and "Garbage in; garbage out." Any thoughtful activity relating to system behavior is limited by the same state of our knowledge. These objections are actually arguments for improving our understanding of the system, improving the models, and improving the data required by the models.

Given a concept of how a system works and given a set of data on the system, there is no better way to describe system behavior than with a computer model. All other activities will underutilize the available resources. The only way to improve the system description is to improve the concept of how the system works and improve the system data set. Modeling is a persistent advocate for both of these improvements.

Postscript: The Grand Global Experiment

Venus and Mars are planets whose climates have evolved to extreme limits. Even though Venus has the highest albedo of all the inner planets, reflecting 71% of the solar energy falling on it, the planet has a surface temperature of 480° C. (Earth averages 15° C.) Venus has a thick atmosphere containing mostly carbon dioxide, and the greenhouse warming of this atmosphere has increased the planet's surface temperature 509° C! Mars, on the other hand, has an albedo of only 17% but a very cold average surface temperature of -60° C, because its atmosphere is extremely thin and produces no observed greenhouse warming. But the Earth's temperature has remained remarkably moderate throughout its entire history (~15° C ± 7° C). Climatologists estimate that the Earth's average temperature changed less than 15° C between the warmest period, the Cretaceous, and the coldest ice age of the Pleistocene.

Even during the early history of the solar system, when the Sun's output was 20% below today's value, the Earth did not freeze over but maintained a moderate climate. Only a planetary negative feedback process could maintain such steady conditions when the system drivers such as solar energy and ice albedo were attempting to push the climate to extremes; most evidence now points to atmospheric carbon dioxide as the mechanism by which Earth has applied the controlling negative feedback.

The fossil fuels that we are now using to provide our energy and are releasing into the atmosphere as carbon dioxide were extracted from the atmosphere and buried during the Cretaceous by Earth system processes. The cooling at the end of the Cretaceous in response to the decrease in atmospheric carbon dioxide corresponds to the demise of the dinosaurs and the emergence of mammals as the predominant complex life form on the Earth's surface. A warming of 6°–10° C will return the Earth to the ice-free conditions of the Cretaceous. Climate models forecast a global warming in the range 1.9°–5.2° C in the next century with a doubling of atmospheric carbon dioxide from fossil fuel use.

All of the consequences of this warming are not known, but based upon the history of this planet the swings in climate of this magnitude have produced major changes in regional and planetary geology, ecology, and biology. Do we really want to perform this global environmental experiment to discover the outcome?

GLOSSARY

absolute zero—On the Kelvin temperature scale there is a coldest temperature called absolute zero, where all motion stops, all molecules are stationary, and all electrons are in their ground state (lowest energy level). This occurs at -273.16° on the Celsius temperature scale.

albedo—The amount of incident radiation that a surface reflects and that thus does not contribute to its heating. The albedo of the whole Earth is approximately 30%. The albedo of snow is ~90% and water is ~10%.

amplify—A system component or process that increases the extent of the fluctuations or changes of a system variable is said to amplify the variable. An audio amplifier takes the small voltage signal produced by the tape player and amplifies it to a large acoustic signal through the audio system speakers. See also **damp**.

annual births—The number of live births occurring in a specified population in one year.

annual deaths—The total number of deaths occurring in a specified population in one year.

asymptotic—A function that gradually approaches a constant value and thereafter remains essentially equal to the constant is said to be asymptotic to the constant.

blackbody radiation—A material that absorbs all incident radiation at all wavelengths appears black and is called a blackbody. Such materials also emit electromagnetic radiation in a predictable manner. Blackbody radiators are the most efficient thermal radiators possible.

branches and decisions—When a model of a system is required to choose among alternate values, paths, or processes, we must supply the system with the required logic and information to make the correct decision; these points in the system diagram are called branches.

cell—The smallest area or volume that is defined in a model is a cell. Only one value of each system variable can be assigned to each cell. For example, in a climate model each cell has one value for temperature for each time step in the model.

climate—The characteristic long-term environmental conditions in a region. Climate can also refer to the whole Earth, as with ice ages and interglacials.

closed loop—If a system variable (or flow) ultimately returns to the reservoir from which it came, then the variable is a closed loop in the system.

computer model—A construct of mathematical and logical statements designed to describe a complex system in quantitative terms with the help of a computer.

conceptual model—A mental image of an object, system, or process.

connector—See **interconnections and coupling**.

conservation of energy—Energy may have several forms (thermal, electromagnetic, chemical, electrical, etc.). Within a closed system, energy may be transformed from one form to another but the total amount of energy cannot change. If the system exchanges energy with the outside world, the change in the total system energy must exactly equal the energy exchanged.

constant— A quantity that has a fixed mathematical value within a model, such as pi or Earth's radius. It may appear explicitly in equations, logical statements, or as an initial value, or it may be a named parameter that the model uses.

constant growth rate—The growth rate (usually expressed as a percent) is the increase in the value of a variable during a specified time period divided by the value of the variable at the beginning of the period. If the growth rate of a variable is a constant, the variable will exhibit exponential growth.

constraints—Restrictions to the behavior of a variable. Constraints are similar to **limits**, but usually act over a broader range of variable values. For example, the global fossil fuel supply has a limit, but prior to reaching that limit national interests, economics, and cost of recovery will constrain its use.

control—See **valve**.

converter—A free-floating element on a system diagram within which new variables or constants may be defined, computations performed, or decisions made.

coupling—See **interconnections and coupling**.

Cretaceous—The geologic period at the end of the age of the dinosaurs, from about 65 to 145 million years ago. During the Cretaceous the Earth was warmer than at present and sea level was higher.

crude birth rate—The number of live births for each thousand persons in a specified population in a given year.

crude death rate—The number of deaths for each thousand persons in a specified population in a given year.

damp—A system component or process that acts to decrease the extent of fluctuations or changes in a system variable is said to damp the variable. By analogy, shock absorbers on cars damp the bouncing that occurs after hitting a bump or dip in the road.

decisions—See **branches**.

demography—The study of human population, the characteristics and dynamics of the whole population and of segments within it.

doubling time—The time required for a function or variable to increase its value by a factor of two.

driver or forcing function—A parameter that controls the behavior of a system and makes its behavior regular and predictable.

dynamic parameter—A value provided to a system that may change with time either in a prescribed manner or in response to the changing state of the system.

e-folding—An exponential function changes by the factor e or 1/e when the exponent of e changes by ±1.0. The time required for the exponent to change, or the distance over which it changes, by ±1.0 is called respectively the e-folding time or the e-folding distance. See also **exponential**.

effective planetary temperature—The temperature that a planet must have in order to radiate to space, as a blackbody, all of the solar power that it absorbs.

electromagnetic spectrum—The entire range of radiation. The wavelengths (distance between adjacent peaks) of the electromagnetic waves within the electromagnetic spectrum range from kilometers, for radio waves, to nanometers (billionths of a meter), for X rays. Visible light is the group of electromagnetic waves with wavelengths between 0.4 and 0.7 micrometers (millionths of a meter).

entropy—The scientific measure of the disorder in a system; the greater the disorder the greater is the entropy. According to the second law of thermodynamics, entropy is always increasing.

equation—A mathematical statement in which equal values (or the mathematical statements producing the values) appear to the right and left of an equal sign. In system modeling and programming, an equation is an action by which the left-hand side is set equal to the value produced by the evaluation of the right-hand side. The two sides are equal after the action is taken but may not have been equal prior to the action.

exponential function—A mathematical function, $y = e^{ax}$ (a is a constant and e is a constant equal to 2.71828), in which y asymptotically approaches 1.0 as ax approaches zero ($ax \ll 1.0$), and y increases without an upper limit as ax increases far beyond 1.0 ($ax \gg 1.0$).

feedback—When information on a system's behavior is used by the system to modify its behavior, the process of transferring the information across the model is called feedback. With **negative feedback,** the modification is in the opposite direction to the behavior and acts as a constraint. In **positive feedback,** the modification is in the same direction as the behavior and so enhances it.

first law of thermodynamics—See **conservation of energy**.

flow or flux—The rate at which a variable enters or leaves a reservoir. By analogy, water in streams flows into or out of reservoirs.

fluctuations—Fluctuations are variations in the value of a variable, usually around the variable's locally averaged value. Hourly temperatures represent fluctuations of temperature relative to the seasonal mean temperature.

forcing function—See **driver**.

forecast—A skilled estimate of some future condition, often based upon computation; by contrast, a prediction may be nothing more than a wild guess.

general circulation model, or GCM—A large three-dimensional computer model of climate that requires a supercomputer. GCMs are uniquely capable of computing the global winds or atmospheric circulation from basic physical equations. GCMs also provide large-scale regional spatial resolution.

ghosted—A system element within STELLA®II that has been duplicated and placed somewhere else in the system diagram; the duplicated component always takes the same value as the original and appears gray, rather than black, on the diagram.

greenhouse gas—An atmospheric gas that absorbs (and also radiates) in the infrared part of the electromagnetic spectrum. Greenhouse gases include carbon dioxide, water vapor, and methane. They warm the atmosphere and surface below them, a phenomenon frequently referred to as **greenhouse warming.**

growth rate—See **constant growth rate**.

guesstimate—An educated guess. In the absence of a reliable value for a needed parameter, modelers will often guess at a value using available information and a hypothesis relating what is known to what is needed.

heat capacity—The amount of heat needed to cause a unit temperature rise in a given mass.

icon—A highly stylized model or image of an object or process (behavior).

if statement—A decision-making instruction common to many programming languages. For example: IF "rain" THEN "dinner" ELSE "picnic," where "rain," "dinner," and "picnic" are variables in the model.

infrared radiation. The region of the electromagnetic spectrum with wavelengths longer than visible light (~1 μm) but shorter than microwaves (~1 mm). The Earth's radiation into space is predominately infrared radiation.

initial conditions—When we run a computer model we give it a "start time," which is the time or date for the computations to begin; the initial conditions are the values assigned to all of the model variables at the "start time."

interconnections and coupling—Parts of a system are coupled if information from one part is provided to, and influences the behavior of, other parts. The information being passed is the interconnection. On the system diagram the interconnection is called a connector.

isolated system—A system that has no significant interactions with other systems or with the rest of the universe.

limits—Most real systems have limits on the range of values permitted for system variables. For example, a power plant's maximum output is limited by its physical capacity, and its normal output is determined by consumer demand.

linear—Linear means, literally, in a straight line; a linear function will plot as a straight line on a graph. If y is a function of x, the equation for the linear function y is $y = ax + b$, where a and b are constants. See also **quadratic**.

main program—See **program**.

model—An idea of an object, system, or process. We may translate this idea into a physical object, a drawing, a mathematical expression, a computer program, or another representation.

model verification—A test or series of tests to compare a model's output with known results; it is an important step to prove the model's accuracy.

negative feedback—See **feedback**.

noise—Unwanted fluctuations in a variable that represent some system process that is not being studied. By analogy, acoustic noise is an unwanted component of sound.

normalized function—A function is normalized by multiplying or dividing all of its values by the same constant. One purpose is to allow different functions to be plotted on the same graph for comparison.

parameter and variable—A parameter is a numerical value that supplies a system model with quantitative information about the system; a variable is a dynamic property of the system that describes the system's behavior. The radius of the Earth and the human population in 1990 are parameters; the mean temperature of the Earth and the current human population are variables.

phase shift—When two related events occur at the same time or two related variables change at the same time, we say that they are "in phase." If one event occurs after the other or the variable change of one occurs after the other we say that a phase shift has occurred.

Pleistocene—The geologic epoch from 10,000 to 1.6 million years ago, which immediately precedes the present epoch (Holocene); the Pleistocene epoch was noted for periodic ice ages.

positive feedback—See **feedback.**

pressure—Pressure is the force per unit area exerted by the collisions of randomly moving molecules.

program—A complete sequence of instructions written in a programming language that the computer can understand. The program must contain all of the information and logic that the modeler wants incorporated in the model. The computer cannot invent information or behavior; all information must be provided in the program and all allowed behaviors described. The **main program** is the part of a computer program that contains the fundamental logic and branches of the model; it always contains the starting point

and the normal ending point. Many of the detailed computations and decisions are made in separate parts of the computer program called **subroutines.** A program can have only one main program but may have many subroutines.

quadratic function—In a quadratic function the highest power of the independent variable is 2. If y is a quadratic function of the independent variable x, the general algebraic form for the function is $y = ax^2 + bx + c$, where a, b, and c are constants. See also **linear**.

regulator—See **valve**.

reservoir or stock—A component of a system that can store or accumulate a quantity of one of the system variables and/or can act as a source of that variable. By analogy, a water reservoir stores the stream water feeding it and supplies water to users downstream.

resonance—Some systems oscillate with a period equal to or related to the system time constant. If a system driver is acting at one of these periods the system response will be amplified with respect to its response at other periods; this condition is called resonance. Striking a bell initially excites many periods or frequencies of vibration in the bell; however, only those periods or frequencies that are in resonance with the bell's natural time constant or frequency will linger to contribute to the bell's continuing tone. See also **amplify**.

run—A completed computer model or program is "run" on a computer. The program or "code" is loaded or stored in the computer's memory, and the computer follows the instructions encoded in the program. Once started the computer proceeds to perform the programmed tasks without external intervention.

second law of thermodynamics—See **entropy**.

sensitivity study—A modeling study to evaluate the magnitude of a change in model output for a given change in input to a specified parameter.

SI units—The Système Internationale d'Unités, the internationally endorsed form of the metric system.

sink—A reservoir that receives a variable from the system under consideration. Usually sinks are large reservoirs that are unaffected by the system being modeled.

solar constant—The amount of solar radiant energy received at the top of the Earth's atmosphere per second per unit area when the Earth is at its average distance from the Sun. The solar constant is 1,368 Wm^{-2}.

solar radiation—The electromagnetic radiation emitted by the Sun. The spectral region from ultraviolet through infrared is important to Earth's climate and weather. See also **electromagnetic spectrum**.

source—A reservoir that supplies a variable to a system. Like sinks, sources are usually large reservoirs that are unaffected by the system being modeled.

specific heat capacity—The heat capacity of a homogeneous substance divided by its mass.

steady-state solution—The final and unchanging result obtained after a model proceeds through early changes. Models that produce steady-state solutions are called steady-state models. They usually feature some form of negative feedback.

Stefan-Boltzmann constant—The Stefan-Boltzmann constant, σ, appears in radiation equations. $\sigma = 5.67 \times 10^{-8}$ $Wm^{-2}K^{-4}$. See **black-body radiation**.

STELLA® II—a graphics-based programming system for the Apple Macintosh™ and Windows systems that allows the user to create computer models without having to write a program in a programming language. It is available from High Performance Systems, Inc., 45 Lyme Road, Hanover, NH 03755 (telephone 800-332-1202). STELLA® II is a registered trademark of High Performance Systems, Inc.

stock—See **reservoir**.

subroutine—See **program**.

supercomputers—Special computers that are designed to solve problems requiring very large arrays of memory and high-speed computational capabilities. Supercomputers are very expensive, and their use is usually restricted to problems requiring their special capabilities.

swamp model—A climate model in which the Earth's surface is treated as a stationary layer of water.

system—A selected set of interacting components usually small enough that its behavior can be understood or modeled. A simple system is the air conditioning system in your home; the global climate system is a complex system. A **system diagram** uses graphic symbols or icons to represent system components in a depiction of how the system works. A **system model** defines all of the interactions among the components of a system and the significant interactions between the system and the outside universe.

thermal energy—The form of energy expressed in the random motion of molecules. When the thermal energy is increased, the molecules move faster and the temperature increases.

thermodynamics—The science that focuses on the flow of energy into and out of systems, the conversions of energy forms within systems, and the influence of the system's energy on its variables.

threshold—A threshold value for a variable represents a value of the variable that separates two usually markedly different behaviors of the system. The system behaves one way below the threshold and a different way above the threshold.

time constant or time scale—The time required for a specific process to occur or for a variable to change significantly. The time constant relating a reservoir to a connected flow is the reservoir divided by the flow.

time step—A computer model progresses in time steps (a defined period of time such as one second or one year) by computing the changes that will occur in all of the model variables during the period of the time step.

tropopause—The level in the atmosphere that marks the top of the troposphere; it is the altitude at which the temperature no longer decreases with height. See also **troposphere**.

troposphere—The lowest layer of the Earth's atmosphere, characterized by the decrease in temperature with altitude. On the average the thickness of the troposphere is about 12 kilometers, but it varies from about 8 kilometers in the polar regions to 15 kilometers in the tropics.

valve or control or regulator—In modeling, the mechanism that specifies the flow through a specific path in a system. By analogy, a valve controls the flow of water from a faucet.

variable—See **parameter and variable.**

weather—The condition of the atmosphere at a given time and place; weather is described by variables such as temperature, wind direction and speed, cloudiness, and precipitation.

work—In physics, a force does work on an object (system), when the object moves or changes its dimension in response to the force. The energy increase of the object equals the work done by the force.

work-by and work-on—The *by* and *on* tell us which direction the work energy is flowing. *Work by* involves energy leaving the system; *work on* involves energy entering the system. If a system changes its dimensions in response to an *internal force*, the *work* done *by* the system is positive. If a system changes its position or dimensions in response to an *external force*, the *work* done *on* the system is positive.

ADDITIONAL READING MATERIAL

Forrester, Jay W. *World Dynamics*. Cambridge Mass.: Wright-Allen Press, 1971. This excellent book utilizes the system diagram and system dynamics approach to elucidate the interactive nature of Earth systems. The authors of the STELLA® II software credit Forrester as "the creator of the conceptual framework and methodology which has come to be known as 'system dynamics.'"

Hannon, Bruce and Matthias, Ruth. *Dynamic Modeling*. New York: Springer-Verlag, 1994. This text is for general college students and is entirely devoted to creating dynamic models using STELLA. Economic, ecological, and engineering systems are considered.

Hutzinger, O., Ed. *The Natural Environment and the Biogeochemical Cycles*. Volume 1 Part E of *The Handbook of Environmental Chemistry*. Berlin: Springer-Verlag, 1990. For the advanced science student. There are four chapters that explore the Earth and ecosystems from the systems point of view.

Luenberger, David G. *Introduction to Dynamic Systems: Theory, Models, and Applications*. New York: John Wiley & Sons, 1979. For math, science, and engineering students in upper-level undergraduate studies. The subject of system dynamics is presented with a broad overview of its wide application to many disciplines.

Meadows, Dennis L.; William W. Behrens, III; Donella H. Meadows; Roger F. Naill; Jørgen Randers; and Erich K. O. Zahn. *Dynamics of Growth in a Finite World*. Cambridge, Mass.: MIT Press, 1974. This book provides the technical documentation for the "World3" global model, the first published version of the Club of Rome-sponsored project to create computer global models.

Meadows, Donella H., John Richardson, and Gerhart Bruckmann. *Groping in the Dark: The First Decade of Global Modeling*. New York: John Wiley & Sons, 1982. This book contains, but is not limited to, the Conference Proceedings of the Sixth IIASA Symposium on Global Modeling (IIASA = International Institute for Applied Systems Analysis in Laxenburg, Austria). This book is unlike any conference proceedings you have ever seen; humor is actually incorporated in a scholarly publication.

——, Dennis L. Meadows, Jørgen Randers, and William W. Behrens, III. *The Limits to Growth*, New York: Signet Books, 1974. The layperson's version of the "World3" global model (see above).

Office of Technology Assessment. *Global Models, World Futures, and Public Policy: A Critique*. Washington, D.C.: U.S. Government Printing Office, 1982. Reviews and critiques various global models, and provides comparative information on their forecasts. The book also candidly discusses the limits to government foresight and the ability of governments to utilize long-range forecasts and planning.

Peterson, Steve, and Richmond, Barry. *STELLA® II Technical Documentation*. Hanover, New

Hampshire: High Performance Systems, Inc., 1994. As the title says, this is the technical documentation, the fundamental reference book on STELLA II.

Richardson Jacques, Ed. *Models of Reality: Shaping Thought and Action.* Mt. Airy, Md.: Lomond Publications, 1984. A marvelous collection of 21 chapters by recognized international scholars of global systems. Several were previously published by the United Nations Educational Scientific and Cultural Organization (UNESCO).

Richmond, Barry. *Authoring Module: Authoring Software for use with STELLA® II and ithink®.* Hanover, New Hampshire: High Performance Systems, Inc., 1994. The latest versions of STELLA II (3.0.2 and higher) have an authoring "shell." This allows an author using special software to create a model to be run by others, who may alter model parameters specified by the author but may not otherwise change the model itself.

Richmond, Barry, and Steve Peterson. *STELLA® II An Introduction to Systems Thinking.* Hanover, New Hampshire: High Performance Systems, 1992. This volume is much more than a manual for the STELLA® II software; there are sections on systems thinking, the principles of modeling, and system behavior.

——, ——, and Peter Vescuso. *STELLA® II Applications.* Hanover, New Hampshire: High Performance Systems, 1992. There are many examples of modeling from psychology, biology, economics, physics, management, etc.

Soltzberg, Leonard J. *The Dynamic Environment, Computer Models to Accompany Consider a Spherical Cow.* Sausalito, California: University Science Books, 1996. Environmental science is transformed into an interactive experience with this product that combines the highly acclaimed book *Consider a Spherical Cow* with the powerful demo version of STELLA® II software.

APPENDIX
*Getting Started With STELLA® II: A Hands-On Experience**

Contents

Introduction

The STELLA® II Demo disks that are provided with this edition of "System Behavior and System Modeling" include new features of STELLA II 3.0 that are not used in your "System Behavior and System Modeling" text. This latest version of STELLA II has a "High Level Map," which is an organizational feature that supports building complex systems in a modular format. In addition, the latest version of STELLA II offers an Authoring add-on, which enables the model builder to transform a completed model into an interactive, simulation-based, gaming environment. Both high level mapping and authoring capabilities are incorporated in the demo disks. These features are described in this appendix, "Getting Started With STELLA® II: A Hands-On Experience." You will probably want to explore all of the capabilities of STELLA II 3.0. However, neither high level maps nor authoring capabilities are essential to building simple system models.

The System Diagram that we describe in "System Behavior and System Modeling" is the middle level, described as the "Diagram Level" in the appendix. On the left-hand bar of the STELLA II display, you will see a small box with up and down arrowheads. You use these arrow buttons to change levels. The "High Level Map" is on the top. The "System Diagram" is on the middle level. The "Equations Listing" is on the bottom level. The middle level, or System Diagram, is where the model building actually takes place. If you wish to bypass the High Level Map and go directly to the simple model building activity, then use the arrows to move to the System Diagram level. If the Globe icon appears in the left bar once you are on the System Diagram level, you should click on it to change from the mapping mode to the modeling mode (indicated by X2). You may now proceed with building your own dynamic models.

You should study Sections 2, 3, 4, and 6 in this appendix to learn more about creating dynamic models. You will find the Quick Help Guide (Section 6) to be especially helpful.

1. Working With A Learning Environment

Welcome to the **STELLA II** Tutorial! The **STELLA II** software is designed to help people build their understanding of dynamic systems and processes. In what follows, you will find an example of a Learning Environment based on a simple **STELLA II** model of population dynamics. We encourage you to spend some time exercising this Environment so you can get a feeling for **STELLA II** as a tool for facilitating learner-directed learning. In the second section of the tutorial, we will take you step by step through the process of building a model. After completing construction of the base model, we will give you an opportunity to extend the model, or you can try your hand at constructing a model of your own choosing.

This tutorial assumes you are familiar with either the Macintosh® or Windows™ environments. If you lack familiarity with the operating system on your machine, you will want to review the User's Guides and tutorials provided with your Macintosh or with Windows.

arning *vironment* *ntext*

Population dynamics is a key content area in many social and physical science curricula. From one-celled organisms to human populations, a generic structure can be used as a basis for studying the dynamics of population growth and decline. In this tutorial, we will examine a deer population in a forest ecosystem. The deer population is regulated by food supply and a predator population. In the year 1900, in response to pressure from local ranchers and farmers, a $50 bounty was placed on predators because of attacks on livestock. In the ensuing forty years, an overshoot and collapse pattern emerged in the deer population, as seen below.

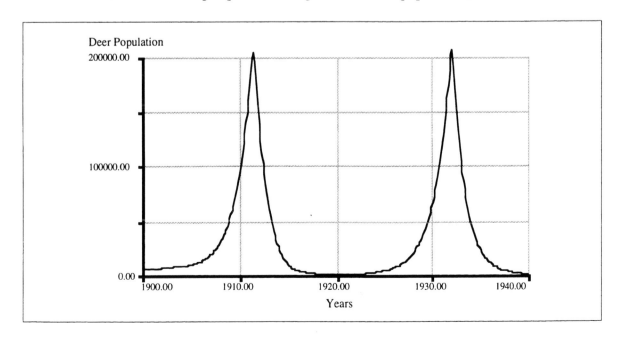

59

The deer population grew until it was well in excess of what its food supply would support. Starvation on a massive scale then occurred. This pattern is frequently observed in the natural world. Your challenge is to intervene in the system in such a way as to establish and maintain a balance between the three members of this simple ecosystem. In doing so, you should try to understand what factors are creating the overshoot and collapse, and determine what points of intervention will have the most leverage in re-establishing a balance.

If you have not yet installed the **STELLA II** disks, do so now. Then locate the folder called *"Tutorial"* and open the model called *"PopDynam."*

What you should see is the High-level Map shown below:

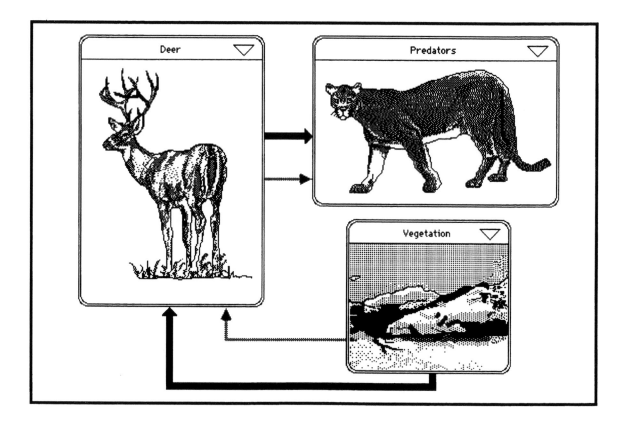

Learning Environment Overview Each process frame in the High-level Map represents a key "actor" in the ecosystem: *Deer*, *Vegetation* and *Predators*. Notice that there are some arrows linking the frames. These show that relationships exist between them, which makes sense since none of these actors exists independently. All three depend either directly, or indirectly, upon each other.

Now, scroll down the page until your screen looks like the picture at the top of the next page.

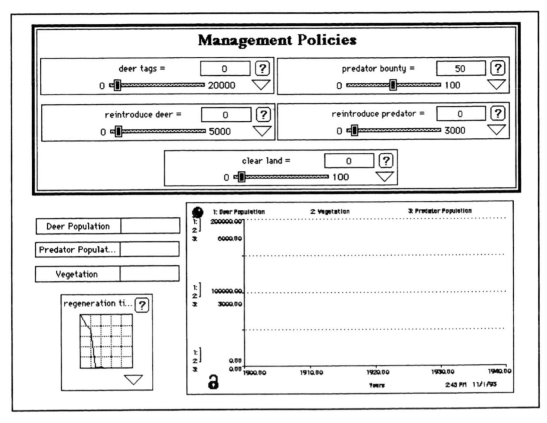

These are the input and output devices associated with this Learning Environment. Using them, you will be able to "fly" the ecosystem and to test your ability to take actions that will keep the system in balance. Along the middle of the left side of the screen you will find three numeric display devices. There is one representing the magnitude of each population in the ecosystem. You will be able to see the current value of population for each actor throughout the course of the simulation. To the right, you will see a graph which is set up to show you the trend traced by each of the three populations as the simulation unfolds. Notice the "push pin" in the top left corner of the graph. If you click once on the pin, the graph will "un-stick" from the control panel and become a separate window. While it is unpinned, you can reposition it or resize it (using the re-size box in the bottom right hand corner) if you can't read the details at the current size. Be sure to pin it back down before clicking elsewhere on the control panel or the window will move behind the control panel when you click on it. (If this does happen, double-click on the small square icon labeled *Graph Output* on the control panel, and the graph will reappear.) Tables and graphs provide the most frequently used ways to view the output of a model. Other alternatives include the numeric display, simple diagram animation and QuickTime™ movies. (QuickTime is only available in the Macintosh version.)

Above the graph and numeric displays, you will see five slider input devices. These devices are directly linked to a corresponding model variable and will allow you to change the values of these variables when you "Run" the model. Click-and-hold on the "knob" in the *deer tags* slider and slide it back and forth. As you slide the knob, notice that the value in the box above the slider changes – this changes the current deer tag policy (the number issued). Alternately, you can type a number into the box which displays the current value of the slider. Click on the Restore button to return the slider to its original value. The final device, in the lower left corner, is the graphical function input display. This device is set up to show the regeneration time required by the vegetation. We will go into more detail about it later.

Simulating:
The Base
Case

We're ready to run our first simulation. Let's see what happens! Make sure all the sliders, the numeric displays and the graph pad are showing on your screen (make sure the graph pad is pinned). Click-and-hold on Restore under the Map & I/O Menu, then choose All Devices in order to ensure that the model is set up with a clean slate. Then, click on the Run icon ($\mathbf{\dot{\lambda}}$) in the lower left corner of the screen. The Run Controller (to the right) will appear. (To move the Run Controller, click-and-hold on the top border then move it to the desired location.) Click on the Run button (or you can choose Run from the Run Menu). As the model runs, you will see the simulation time below the Buttons. The graph will show what the three populations do over forty years with the predator bounty in effect.

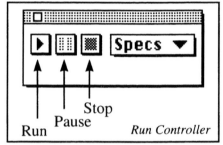

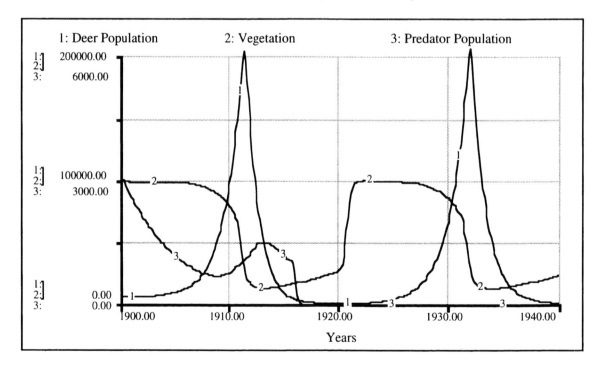

You can see that removing predators from the system in 1900 sets into motion an exponential growth pattern in the deer population. Even with the abundant prey provided by such a large deer population, the predators were not able to survive the impact of the bounty. Also notice that the vegetation, once grazed down, recovers very slowly at first. The model is set up to reflect the effect of overgrazing on the vegetation regeneration time. When plants are eaten back to a point where their root systems are damaged, they recover at a very slow rate. Once they restore their growth infrastructure, the growth rate returns to normal.

You can change this variable regeneration time in the graphical function input display (lower left corner of the control panel). Double-click anywhere on the graphical function input display. A dialog will open up revealing a graph that shows the relationship between *Vegetation,* relative to its initial value at the beginning of the simulation, on the X axis, and *regeneration time,* on the Y axis. The model calculates how much *Vegetation* is present during the current year compared to the initial value, and then chooses the corresponding value for *regeneration time.* As long as this calculation stays above .7, *regeneration* continues at the normal rate. You can change the pattern of the curve, and thus the regeneration time, by placing the cursor inside the grid (the arrow will turn into a cross-hair). Click-and-hold, then move the cursor around. The curve will follow the movements of the cursor. Click OK and Run the model again. What effect does the change you made in *regeneration time* have on the model behavior? Does a faster or slower growing plant impact the dynamics? When you are done with your experiments, click the Restore button to return the original curve to the graph.

Policy Interventions

Now is your chance to try your hand at managing the system to improve its performance. You will have five options at your disposal:

1. *deer tags* allows you to issue or restrict hunting licenses in an effort to control the herd size.
2. *reintroduce deer* allows you to bring in more deer, should their numbers drop below what you consider to be desirable.
3. *predator bounty* allows you to increase or decrease the amount of the bounty; the higher the bounty, the larger the hunter population out there trying to shoot predators.
4. *reintroduce predator* allows you to bring in more predators, should their numbers drop below what you consider to be desirable.
5. *clear land* allows you to see the impact of changing land use in the ecosystem (land is measured in hectares; there are 1000 hectares at the beginning of the simulation).

If you have any questions about what each slider represents, click on the "?" on the slider. This will open up a document field that will give you additional information. You will find these question marks on many of the elements on the control panel so make full use of them.

Just a reminder: Your challenge is to intervene in the system in such a way as to establish and maintain a balance between the various members of the ecosystem. In doing so, you should try to understand what relationships are creating the problem, and what points of intervention have the most leverage for establishing an appropriate balance. Don't just react! See how well you can do at achieving a balance.

Before you try your hand at improving things, let's make a quick change in the way the model runs. Go to the Run Menu and choose Time Specs. Change the value of Pause Interval from INF (Windows) or ∞ (Macintosh) to 4. This will cause the model to pause every four years, allowing you to make the changes you think will bring the system into balance. You also can pause any time you want by clicking on the Pause button on the Run Controller, choosing Pause under the Run Menu or clicking-and-holding on a slider knob. Click OK.

Make any changes you'd like to the five policy variables. Then, click the Run button to begin the simulation. When the model pauses, make any changes you wish to the values of the sliders. When you are done making your changes, click on the Run button. Alternately, you can select Resume from the Run menu.

A Last Look from the High-level Map

Now, one last look from the High-level...

Use the scroll bar on the right side of the page to take you up to the very top, so you see the map which appears on page 2. Click once on the downward pointing arrow on the top right side of the *Deer* process frame.

Surprise! We've "lifted the hood." You are now looking at the engine that has created all the dynamic behavior you've observed. Scroll around and explore the model. In the next section, we are going to show you how to create some of this engine. If you want to go back up to the High-level, click the arrow on the sector header (it looks like the one you clicked to get down here except it points upward) and you'll be there. Alternately, you can click the upward pointing arrow on the left border and you will be moved to the High-level.

Once you have completed your experiments with this ecosystem, move on to the next section. There you will learn how to build a model such as the one used in the preceding exercise, for yourself.

2. Building a Model

Context for the Model

Studies of diverse populations have shown that there are a few common patterns of growth which emerge repeatedly. "Over-shoot and Collapse" (drawn below) is one such pattern. We'll now construct a model of a deer population to see if we can discover what relationships are responsible for generating this pattern.

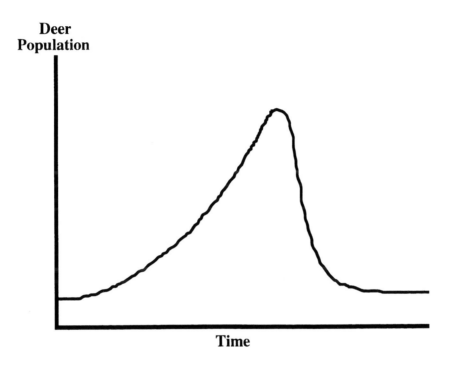

Mapping: High-level

At the beginning of every modeling effort, it is important to focus your efforts by identifying the most important "players" in the system you are interested in understanding. The most important players in the system you experimented with in the previous section are deer, predators and vegetation. We will begin by laying out a High-level Map showing these players. Then, we will navigate to the Diagram Layer to build a portion of the related model. **We have provided a Quick Help Guide, beginning on page 88 of this booklet, which details the mechanics of model building. Please refer to this guide if you encounter problems in your modeling efforts.**

If you still have **STELLA II** open to the "*PopDynam*" model from the previous exercise, choose Close Model from the File menu. Save changes if you wish. Then, choose New from the File menu. If you are just starting, double-click on the **STELLA II** program icon. A blank page will open up.

Using the process frame building block, left-most on the palette, begin by creating the High-level Map shown at the top of the next page. *(Hint: If you hold down the alt key (Windows) or the option key (Macintosh), the hand will turn into the last*

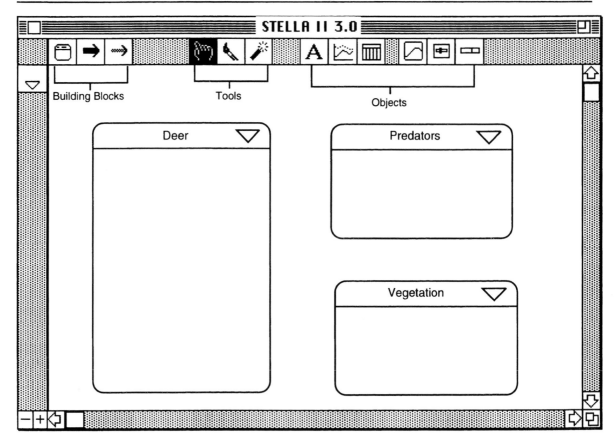

tool used. So if you hold down the alt or option key after placing your first process frame on the screen, you can put down the other two without going up to the palette each time.) If you make a mistake, use the dynamite tool ( ) or choose Clear from the Edit Menu.

Mapping:
Diagram
Level

Now let's get into more detail about two of these actors. Click on the downward-pointing arrow in the header of the *Deer* process frame. This is what a sector looks like on the Diagram Layer. Inside this frame, we will put all of the parts of the model that relate to deer.

Using the building blocks at the far left side of the palette, begin creating the map shown at the top of the next page. **If you encounter any difficulties with your mapping efforts, refer to the Quick Help Guide which begins on page 88 of this booklet.** To move the sector frames, click-and-hold on the header of one of the sectors, then move it to the new position. To resize it, click once to select it (notice the "handles" that appear), then click on one of the handles and drag. The sector will resize. Make sure your flow ( ) is connected to the stock ( ) — there should only be one cloud — and that there are two connectors ( ).

Modeling:
Round 1

Once you've created the simple Population map, we'll next want to bring the map to life through simulation and animation. To do so, we'll need to enter a few numbers and a couple of relationships. But first, you should note that the

66

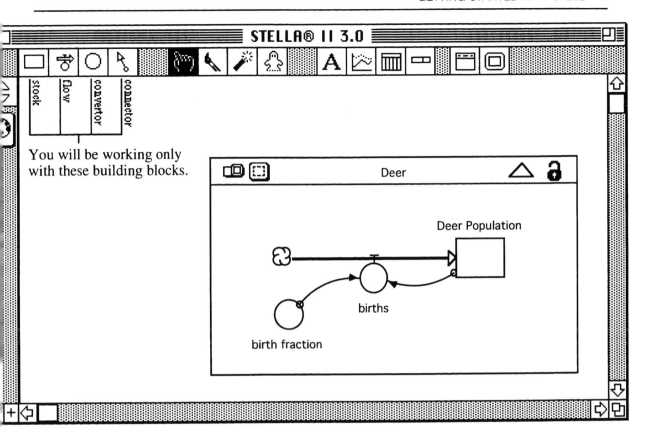

STELLA II software already has given you a big headstart in this process by *automatically* creating the equation framework needed to do the basic "accounting" for the system. This framework is *generic*, and thus the software can automatically create it for any system — no matter what the context. To take a look at the framework, click once on the downward-pointing arrowhead located just above the globe icon on the left border of the diagram frame. Here's what you'll see:

```
Deer
[?]  Deer_Population(t) = Deer_Population(t - dt) + (births) * dt
     INIT Deer_Population = { Place initial value here... }
     INFLOWS:
        births = { Place right hand side of equation here... }
(?)  birth_fraction = { Place right hand side of equation here... }
```

For each accumulation, represented by a rectangle (or "stock"), and the associated set of flows (pipes with arrows), the software creates a generic equation. The equation says, in words: *"What you have now, is what you had an instant ago (i.e., 1 dt in the past), + whatever flowed in over the instant, - whatever flowed out over the instant."* The software automatically assigns + and - signs **based on the direction of the flow arrowheads in relation to the stock.**

Now, in order to simulate, the software needs to know "How much is in each accumulation at the outset of the simulation?" It also needs to know "What is the flow volume for each of the flows?" The answer to the first question is a number. The answer to the second may be a number, or it could be generated by a relationship. We'll illustrate both as we move from the mapping to the modeling phase. Click once on the upward-pointing arrowhead on the left frame of the equations window. This will return you to the diagram. Then, click once on the Globe icon (🌐). Doing so will shift you into the *modeling* mode (the icon will change to an X^2), and a ? will appear in each icon on the diagram. The ?'s indicate the need for information to enable the model to be simulated.

Let's begin by thinking about the dynamics of a population. It makes sense that *births* depend on *Deer Population* since an increase in deer will cause an increase in deer births. In order to show this dependency in the map, we used the Connector (⇲) to link *Deer Population* to *births*.

Double-click on the *births* flow regulator (the circle on the pipe). **Click in** the equation *(Deer Population * birth fraction)* by clicking on the respective variables in the Required Inputs List — *don't type in the variable names (you might make a typo, and* **STELLA II** *makes it unnecessary to repeat this typing, as you've already done it once!).* Click in the * sign linking the two variables from the Calculator key pad in the dialog box; alternately, you can type in the * sign from your keyboard. When you've established the correct equation, click OK to exit the dialog. Note that the ? no longer appears on the *births* flow regulator.

Next, we need to determine an initial value for our *Deer Population* at the beginning of the simulation. We will say it is 100. Double-click on *Deer Population* and click in (or type) "100." Click OK. Now, to define the *birth fraction*, let's assume that there are two births for every 10 deer in the population or a birth fraction of .2. Double-click on the *birth fraction* convertor and click in (or type) .2. Click OK to exit the *birth fraction* dialog.

Simulating:
Round 1

Now, let's run a simulation of this simple model. Click on the Run icon (🏃) to open the Run Controller. We will need a way to view the results. So, let's create a graph. To do so, select the Graph Pad icon from the Objects palette (📈). Click once on the diagram, and you will deposit a graph pad in that spot. Double-click on the resulting blank Graph Pad page. Since *Deer Population* is the variable that we want to track, move *Deer Population* from the Available list to the Selected list within the dialog. **(If you encounter any difficulties defining your graph, refer to the Quick Help Guide which begins on page 88 of this booklet.)** Finally, click OK.

Before we simulate, choose Time Specs... from the Run Menu. Then click the radio button next to "Years" which appears in the list under the heading "Unit of time:" This will cause "Years" to become the time unit on the graph. Click OK.

Run the model and watch the graph of *Deer Population* unfold over time. (If your graph has disappeared behind the diagram, double-click on the Graph 1 icon to bring it forward again.) The scale for Deer Population should run from 100 to 1040.13. This is not a very "pretty" scale to display your results. So, let's specify a scale for the graph, rather than letting the software set it for us.

Double-click on the graph. Click once on *Deer Population* in the Selected list. While *Deer Population* is selected (it should be highlighted), click once on the two-headed arrow (↕) to the right of *Deer Population.* Horizontal lines will appear above and below the arrow (↨). These lines indicate that you are about to set a "local" scale (i.e. a "floor" and "ceiling") for the variable. "Local" means that the scale applies to this graph only. (It is also possible to set a global scale for a variable that will apply to *all* graphs in the document.) Notice the scale boxes below have now become editable. Type "0" in the Min box and "1500" in the Max box. Then click the Set button. Click OK to exit the dialog — you now should have a new scale on your graph.

If you'd like to see actual numbers, you can view them by creating a Table. Double-click (Windows) or click (Macintosh) the close box in the upper left corner of the Graph Pad page to put the page away. Next, select the Table Pad icon from the Objects palette (▥). Plop it down next to the Graph Pad icon. Double-click the resulting page and enter the *births* flow and *Deer Population* stock into the Selected list. Then, click OK, and Run the model. The Table results should show you how **STELLA II** is calculating the values for the entities you selected. Close the table when you are finished with the run.

Before we move to the next step in the modeling process, let's take a quick look at the animation capability within the software. There are two types of animation available within **STELLA II**. The first is animation of diagram icons. The second is the use of QuickTime movies. (The latter is only available on the Macintosh). We will take a peek at diagram animation.

Choose Diagram Prefs... the last item under the Diagram Menu. Under the word "Animate," you'll see three icons. Click once on each. Before clicking OK, peruse the dialog. It's in here that you can add more "pages" to the diagram, change diagram font types, etc. Play, if you wish. Then, click OK. Run the model and watch the diagram elements. You'll see that the *Deer Population* accumulation will fill over time. Little needles will mark the volume of flow in the flow regulator on *births* and in *birth fraction.* This is a simple way to view the patterns generated by the simulation.

Modeling: Round 2

Remember the "overshoot and collapse" dynamic we are seeking to represent? The graph and table results definitely suggest that there is something missing from this picture. *Deer Population* is growing exponentially with nothing to check its progress. There is no collapse to the pattern. In order to generate a collapse, we will need to have some check on *Deer Population* growth. One obvious check is *deaths*. Add a *deaths* flow and a *death fraction* to the diagram to make it look like the map shown below. To enlarge the sector frame, move the hand over the header until it turns into an arrow. Click once. Notice the "handles" that appear in the corners. Click-and-hold any one of the handles and the sector will resize according to where you drag.

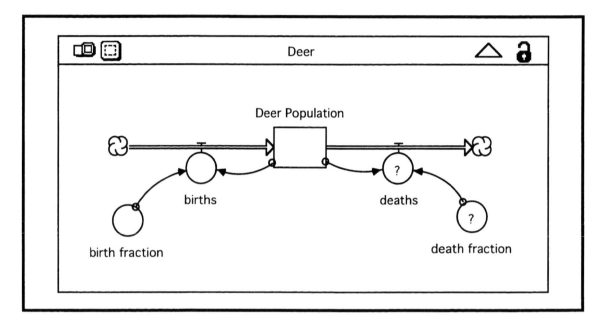

Next we will define the two new entities. Open the *deaths* flow and click in *Deer Population* from the Required Inputs list. Then click OK. You will encounter a message telling you that you have not used one of the Required Inputs: *death fraction*. With **STELLA II**, "What you see is what you get!" If you have shown a relationship in your map, then the relationship must be included in your equation.

Position your cursor in the equation box after *Deer Population* and click in "**death fraction*." Then type in "+*pollution*." Click OK. You should get another message telling you that *pollution* is not an object on the diagram. It is also true in **STELLA II** that: "What you don't see, you *don't* get." If you haven't shown the relationship in the map, you *can't* use it in the equation. Delete the "+" and "pollution." *deaths* should now equal *(Deer Population * death fraction)*. Click OK. Then, set the *death fraction* to .02.

Simulating: Round 2   How do you think the addition of a *deaths* outflow will alter the pattern traced by *Deer Population*? To find out, run the model and observe the new behavior on the graph. What you should have seen is that *Deer Population* continues to grow in an exponential pattern. Even though we have *slowed* the growth, we have yet to top it, or to produce the collapse we are seeking.

Perhaps the problem is that the value of *death fraction* that we've chosen simply is too small. Suppose we want to test various values for *death fraction* to see how they affect the pattern traced by *Deer Population*. A **STELLA II** feature called "Sensitivity Analysis" enables you to examine the sensitivity of model performance to a variation in model parameter values.

Choose Sensi Specs... from the Run Menu. Enter *death fraction* into the Selected list within the dialog. Next, replace the 3 with a 4 in the box labeled "# of Runs" (in the middle left side of the dialog). Then, click once to select *death fraction* in the Selected list. Note that the "Incremental" variation option becomes selected by default. The incremental option means that, in this case, *death fraction* will be varied incrementally between whatever two values you enter into the Start and End boxes in the dialog. Enter the number .02 into the Start box and .1 into the End box. The software will determine 4 values, evenly spaced between (and including) .02 and .1. After entering your Start and End, click the Set button. Then click the Graph button on the left in the dialog. Enter *Deer Population* into the Selected list within the resulting Graph dialog. (Note that the "Comparative" option is checked for Graph Type in the dialog.) Click OK. Run the model and watch the four simulations play out on the graph.

If you would like to see the parameter values that were used to produce the four curves, click the ? on the lower left of the graph pad page.

*Modeling:
Round 3*

Still no collapse. The problem is not *death fraction*. So how can we get the population to collapse? This might be a very good question to put to your students!

The answer is that we need to add some type of relationship that either causes the *death fraction* to rise during the simulation to exceed the *birth fraction,* or causes the *birth fraction* to fall to a level below the *death fraction* as the simulation unfolds.

In the previous exercise, as the sensitivity runs clearly indicate, the closer to .1 the *death fraction* came, the less rapidly *Deer Population* grew. Suppose, our Population model allowed the *death fraction* to increase as the *Deer Population* increased? Doing so would proxy the effect a growing deer herd would put on a fixed food supply, for example. **STELLA II** provides a vehicle for including such relationships. That vehicle is called a "graphical function." Let's incorporate the Deer/death fraction relationship into our model using a graphical function.

Draw a connector ( ) from *Deer Population* to *death fraction*. Then double-click on *death fraction*. Click once on *Deer Population* in the Required Inputs list, and then click once on the button labeled "Become Graph." You will see a blank grid appear. Change the default ranges on the X and Y axes of the grid to reflect the numbers in the graphical function shown on the next page. Place your cursor inside the blank grid until it turns into a cross-hair. Then, click-and-hold and then drag the mouse around the grid. A curve will be traced by following your movements. You can try your hand at sketching in the curve you see below or, alternately, you can type the numbers into the Output column. There is no need to reproduce the exact numbers shown in the illustration below. Just capture the shape of the relationship.

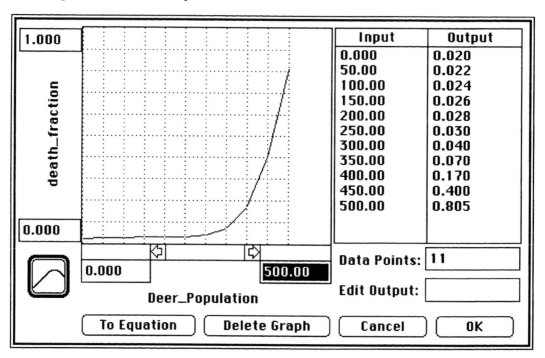

Input	Output
0.000	0.020
50.00	0.022
100.00	0.024
150.00	0.026
200.00	0.028
250.00	0.030
300.00	0.040
350.00	0.070
400.00	0.170
450.00	0.400
500.00	0.805

Data Points: 11

Edit Output:

To Equation   Delete Graph   Cancel   OK

Simulating: Round 3   When you have entered your *death fraction* curve (remember, this relationship is an *input*, not a graph of *output*), click OK. Then, under the Run Menu, choose Sensi Specs... You will get a message telling you the current setup has been changed. Click OK and then click cancel in the dialog box. Now double-click on *Graph 1* to open it and Run the model.

Deer Population no longer grows forever. However, we have yet to achieve "collapse." Instead we've generated a ubiquitous pattern in population dynamics: S-shaped growth! To do so, *death fraction* had to become a variable. You may wish to experiment with alternate patterns for the *death fraction* to see if you can cause the system to generate overshoot and collapse. Can you make this model generate the overshoot and collapse pattern? If not, why not?!

The graphical function affords a lot of power without the need for much mathematical expertise. Many interesting processes across the curriculum contain non-linear relationships. In the past, constructing models which contained such relationships was beyond even the mathematically sophisticated undergraduate student. With **STELLA II**, even students with very little mathematics training will be able to capture non-linear relationships with ease.

As we hope you are discovering, the power and flexibility afforded by the **STELLA II** software is considerable! There really is no end to the interesting experiments that students and/or faculty members might perform with the software. Thus far we have discovered the relationships that create both exponential and S-Shaped growth patterns in a population. But we've yet to produce overshoot and collapse. At this point, can you offer an explanation of what we'd need to do to the model in order to generate such a pattern?

Modeling: Round 4

What we need to do is cause the death fraction to *overshoot* the value of the birth fraction — not just to rise up and equal it. In order for this to happen, the death fraction must depend on some variable other than the *Deer Population*. In reality, this would be true. Deer are not dying because of the size of the Deer Population. They're dying because of some impact their population is having on deer viability. One such impact is on their food supply. Therefore, the next addition to the model will be vegetation.

To incorporate vegetation into our model, we will proceed using the same steps we followed in building the Deer Population. Find the sector frame called *Vegetation*. It will be on your diagram because we put a corresponding process frame on the High-level Mapping Layer in the first part of the modeling section. Move the hand icon over the header until it turns into an arrow. Click-and-hold and drag the sector over until it is below the Deer sector.

Next, put down a stock (☐) inside the sector and name it *Vegetation*. Now we'll need to add two flows ( ) — an inflow, let's call it *regeneration*, and an outflow which we will call *consumption*. Then initialize the stock with 3500 units of vegetation.

How does vegetation regenerate? Regeneration is produced by the vegetation itself. So, we will need a connector ( ) drawn from *Vegetation* to *regeneration*. Each plant produces a certain amount of new growth each time period. This means we will need a converter (◯) called *regeneration per plant*. Use the connector ( ) to show that *regeneration per plant* is an input to *regeneration*. Open *regeneration* by double-clicking on it and specifying the relationship, *Vegetation * regeneration per plant*.

Next, what will generate the *consumption* outflow? The outflow will be generated by the *Deer Population*, who is munching the food. We also need to indicate the amount of food each deer is capable of eating during each time period. Add a converter (◯) and call it *vegetation per deer*. At this point, let's just assume *vegetation per deer* is a constant. Let's set it equal to 15. Connect *Deer Population* and *vegetation per deer* to *consumption* using the connector. Next, make *consumption* equal to *Deer Population * vegetation per deer* and the *regeneration per plant* equal to .5.

The final issue we must address is the effect that vegetation availability has on the Deer's *death fraction*. Using the dynamite tool ( ), "blow up" the connector linking *Deer Population* and *death fraction* (place the fuse of the dynamite inside the connector's take-off button located on the perimeter of the stock).   Next, draw a connector from *Vegetation* to *death fraction* and double-click on *death fraction* to re-define it. You will need to replace *Deer Population* with *Vegetation*. Now click the "To Graph" button, set a new scale and draw in a curve which represents the relationship between *death fraction* and *Vegetation*. It should look something like the curve shown below.

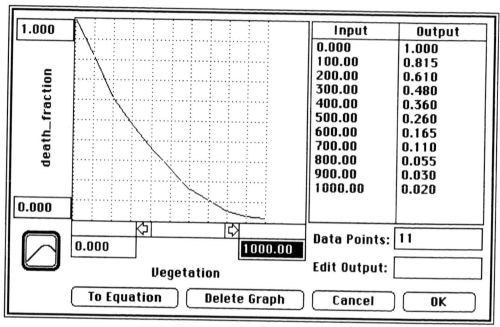

Your model should look like the one shown below (your flows might come from different directions but make sure they are the same relative to the stock).

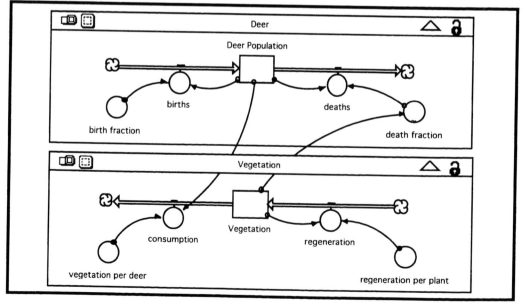

Simulating: Round 4

Now we are ready to run the model. Open your graph pad and examine the behavior of *Deer Population*. Voilà! Overshoot and Collapse! Can you explain why this pattern occurs? You may find it useful to create a graph with the deer *birth fraction* and *death fraction*, scaled on the same axis (see the Quick Help Guide for assistance). Make a run and watch what happens to these two variables as the simulation unfolds.

Final Look at the High-level Map

Let's take one last look up at the High-level Map. Click on the upward-pointing arrow on the *Deer* sector to navigate to the High-level. Notice that the software has automatically created the bundled connectors (the arrows between process frames) that correspond with the structure we put into the model on the Diagram Layer. To rearrange them, click-and-hold on one of the connectors and drag it to a new location. Double-click on one of the connectors. Here you will see the list of connectors between the two process frames that correspond with the direction of the arrow in the bundled connector.

Now click on the downward-pointing arrow on the *Deer* process frame to navigate back to the Diagram Layer. Click-and-hold on the cloud at the end of the *consumption* outflow and drag it up into the *Deer* sector. Click on the upward-pointing arrow again. Now there is a bundled flow on the High-level along with the bundled connectors. (You can double-click on this as well to see what flows are included in it.) This is actually a more accurate picture of the relationships between these two process frames, since the biomass of vegetation is actually being consumed by deer. By putting the cloud in the Deer sector, you will show this relationship on the Diagram Layer and the High-level map.

From this point, you can expand the model to incorporate Predators, climatological effects, or whatever your needs or interests dictate. With a few name changes, this same model could be used to represent immigration into and out of a city or the spread of milfoil in a Northeastern lake. The experiments that are possible with even a simple model such as the one explored in this tutorial are limited only by your imagination and creativity.

When you have finished your experiments with the Deer and Vegetation model, move to the next section where we will show you another interesting feature of **STELLA II**. We are going to be adding a sub-model. Sub-models allow you to add detail to your model without adding too much visual complexity at the same time.

3. Creating Sub-Models

Building a Sub-model Sub-models allow you to "drill down" to view additional model structure while maintaining a simple and neat diagram. We will use this method of managing diagram complexity to add detail to the Deer Population we have been studying. We are going to provide a more detailed representation of the age distribution of the Deer Population.

If you still have the Deer and Vegetation model from the previous exercise open on your machine, close it (save it if you'd like) and choose New from the File Menu.

Click on the downward pointing arrow located on the left border of the High-level Mapping Layer. Doing so will transport you to the Diagram Layer. Now build the simple Deer Population map shown below:

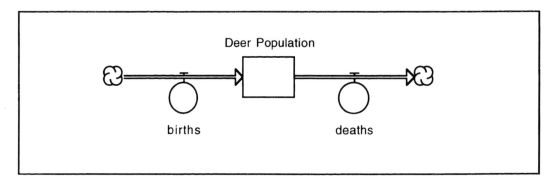

Next, click on the globe icon (🌐) to shift into modeling mode. Then, double-click on *Deer Population*. Notice the choices along the top border of the dialog box: Reservoir, Conveyor, Queue, Oven. Each of these represent a different type of stock. Click once on the button next to the Conveyor. Then, click in the check box next to the word "Sub-model." Finally, click OK. What you now see is the Sub-model icon with an open sub-model space; this is where we will put the drill-down detail associated with the Deer Population.

Before you add the detail, let's look at a couple of issues related to having a sub-model open on your diagram. First, notice how the items on the diagram have become greyed out. This has happened because when a sub-model is open, you are no longer able to directly interact with items on the diagram. In fact, if you move the cursor to a point outside of the sub-model space, it turns into the international prohibition symbol (Ø) to remind you that you are "out of bounds." You can build within the sub-model space using the building blocks in the same way that you did on the diagram. However, if you want to perform operations on the Diagram Layer, you must hold down the Ctrl key (Windows) or the command key (Macintosh) in order to do so. If you need more room in the Sub-Model space, move the Hand until it is over the border of the Sub-model space. When it turns into an arrow, click once. Handles will appear in each of the corners of the space so you can resize it to whatever size you need. If you need to move it, click-and-hold on the border, then drag to move the entire space.

With the sub-model space open, use the building blocks to create a map that looks like the one shown below:

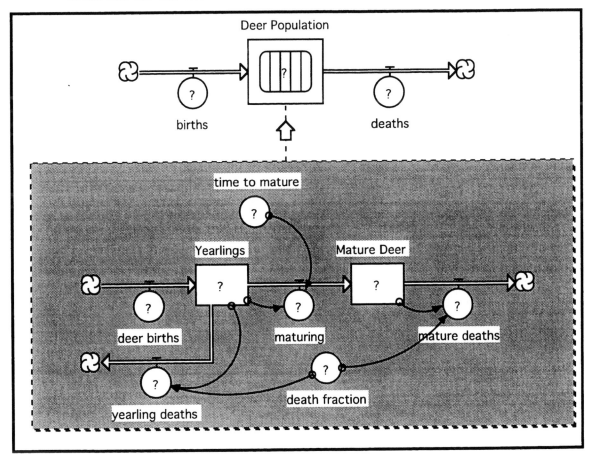

Double-click on the *Yearling* stock and enter 100 for the initial value. Enter 500 for an initial value for the *Mature Deer* stock. Since it is possible to have a great deal of detail in this sub-model, we need to tell the software which chain of activities we want to have totaled and rolled-up into the sub-model icon and associated flow. Hold down the Ctrl or command key and double-click on the sub-model icon (▥). Here you see a list of all the possible choices for roll-up. In our case, there is only one. Click once on the *Yearling + Mature Deer* line. When the software calculates values for *Deer Population*, it will add these two stocks together to yield the number. Click OK.

Next we need to tell the software what flows "roll down" from the sub-model icon inflow to the *Yearlings* inflow. Double-click on the the inflow into *Yearlings*. Notice that *births* is a required input for this inflow. This is because *births* is the *only* inflow to the sub-model icon. Click once on *births*. Then click OK. A special symbol (⇥) has been added to the sub-model inflow and its name has been changed to *births'* to show that it is equal to the *births* inflow which appears on the Diagram Layer.

Next, we need to determine what outflows on the sub-model layer "roll-up" into the outflow called *deaths* on the Diagram Layer. In this case, there are two roll-up outflows since both *Yearlings* and *Mature Deer* die. Hold down the Ctrl or command key and double-click on the *deaths* outflow on the Diagram Layer. Notice that both *yearling deaths* and *mature deaths* appear on the Allowable Inputs list. (If they don't appear on this list, check to be sure the flows are connected to the stocks -- see the Quick Help Guide beginning on page 33 of this tutorial.) Click once on each of them — now they are "rolled-up" into the *deaths* outflow from the *Deer Population*. Click OK. Notice that the names have changed.

Let's define the *death fraction* as .02 and the time to mature as 1 year. Open each of the converters and put these values in.

Now, to define the flow rates. Double-click on the *deaths* flow out of *Mature Deer* and click in the equation *Mature Deer * death fraction*. Click OK. Do the same for the *deaths* flow out of *Yearlings*. Finally, open the *maturing* flow out of *Yearlings*. This one is a little different since *time to mature* is the number of years a yearling stays in the stock rather than a percentage of yearlings maturing each year. The equation to enter into the *maturing* flow to show this relationship is *Yearling / time to mature*.

The last flow is the *births* flow on the Diagram Layer. We need to show a connection from *Mature Deer* to *births*. Use the connector tool (⭭) to do so. The software allows you to make the connection from the sub-model to the Diagram Layer without holding down the Ctrl or command key (since the origination point lies within the sub-model space). Now click the upward-pointing arrow located right below the Sub-model icon. This will retract the sub-model so you can work unrestricted at the Diagram Layer. Put a converter (○) near the *births* flow and call it *birth fraction*. Use a connector (⭭) to show it is an input to *births*. Double-click on *births* and click in the equation *Mature Deer * birth fraction*. Click OK. Set the *birth fraction* equal to .2.

One last tip: Each entity on the diagram has two locations — one when any sub-model is open, and one when all sub-models are closed. We've allowed for a second position so that you can move diagram entities around. This way they won't lie directly under a sub-model space. Open the sub-model again. See if it overlaps your birth fraction converter. With the sub-model open, holding down the the Ctrl or command key, click-and-hold on *birth fraction* and move it to any spot that makes the diagram look neat. Now click on the roll-up arrow at the the top of the sub-model space. When the sub-model is closed, all the diagram entities you moved while it was open should return to where they were before you opened the sub-model.

Simulating

Now you can run the same type of experiments we did in the first part of the tutorial. Or you can try adding a Vegetation Sector. Has the age structure changed the fundamental dynamics generated by the system?

There is another way to view the numbers of the diagram entities as the model runs. With the sub-model closed, click once on the numeric display device in the objects palette (⊏⊐). Click again to place the display below the *Deer Population* sub-model icon (click-and-hold to move it around after you deposit it). Double-click on the numeric display and a dialog will appear that will have a list of all entities in the model. Move *Deer Population* from the Allowable list to the Selected list using the >> button. Click in the box next to Show Name to de-select it. This way you will only see the number not the entity name. Click OK and create any other numeric displays you are interested in seeing. Run the model to see the results.

Space Compression

Another method of managing diagram complexity is the Space Compression Object. If you want to compress a portion of the model without having associated flows to other parts of the model, click once on the Space Compressor (▢) in the objects palette, then click on the diagram. The result will be an object that looks similar to the sub-model from the previous example. From there it will work just like the sub-model space without the "roll-ups" and "roll-downs" to the flows.

The next section of this tutorial will show you two more helpful tools and then a summary of the building blocks, tools and objects and where they can be found in the **STELLA II** *Technical Documentation* for additional information about how they work.

4. Other Helpful Features

The STELLA II software contains many features we have not covered in this tutorial. In this section we will describe two features that can be helpful in adding to the utility and appearance of your model: the text block and the paint brush. Following these descriptions, you will find a summary of all the palette items and where you can find more information about them in the *STELLA II Technical Documentation*.

Text Block   If you want to section off portions of your model with a frame and text (such as the Management Policies frame on our learning environment control panel) or annotate directly on the High-level Map or Diagram Layer, you can use the Text Block object on the object palette. Click once on the text block object (**A**) and click again to place it on the diagram. You will get a block with a simple border. If you want to move (or select) it, move the hand until it crosses the text border; it will turn into an arrow. Click-and-hold and you will be able to move it to any location you choose. To resize the the block, click-and-hold on one of the "handles" (while it is still selected) and drag it to the size you want. To type in the text block, move the hand to the top left hand corner of the block; the hand will turn into an I-beam. Click once and you will be able to start typing. To format the text and change the border, move the hand over the border of the block until it turns into an arrow. Double-click and a format dialog box will appear. Make your choices and click OK.

Using Color   If you have a color machine, you can use the paint brush tool to color any objects or building blocks on the High-level and Diagram Layer, as well as the background page, the graph and table pages. If you click-and-hold on the paint brush tool ( ) in the tool palette, you will see all the colors available to you (this will be determined by what your computer allows as well as the video driver your're using under Windows or the setting you've made in the Monitors Control Panel on the Macintosh). While you are holding your click, drag down and choose the color you want. When you release your click, the paint brush will be the color you selected. Click on the item you want to color. Be careful not to click the surface or you will color it. You can even "tear off" the color palette when you want to color several things all at once. Click-and-hold on the paint brush, drag down off the lower edge of the palette, and the palette will follow your cursor. Now use the palette repeatedly. When you are done, click in the box in the top left hand corner. (Hint: If you want to change the default color for one of the building blocks or objects, choose a color. Then while holding down the alt key [Windows] or option key [Macintosh], click on the item in the palette you want to change.)

Summary of
Palette
Below you will find a summary of the Building Blocks, Tools and Objects and where you will find them documented in the *Technical Documentation*. Note that the last three objects on the Mapping Layer are available only with the Authoring Module and are documented in the Authoring Module Guide. They will be greyed out if you own only the Core version of the software.

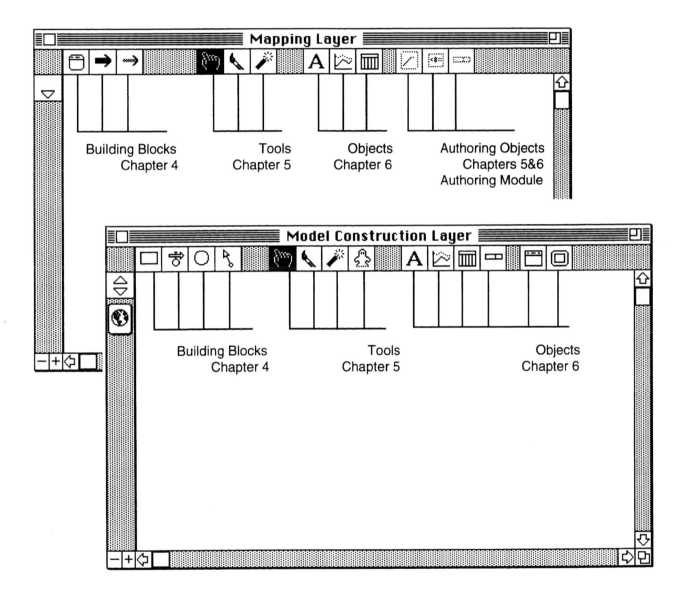

Parting Words
The **STELLA II** software makes it easy (and fun) to build an understanding for complex, dynamic processes. We think you'll find it to be an invaluable addition to your arsenal of learning tools. We have only covered a fraction of the features of the software but the documentation that accompanies the software is designed to help you with many of the challenges you will face as you engage in the modeling of various systems. There are three **STELLA II** User's Guides. The

Introduction to Systems Thinking guide provides further information for building your understanding of the Systems Thinking framework and concepts. The *Applications* guide provides illustrations in various substantive contexts ranging from the social to the physical sciences. Finally, the *Technical Documentation* guide provides detailed, feature by feature descriptions of the software. We encourage you to take full advantage of all that is there. Please feel free to contact us for further information or for technical support as you proceed from here.

If you have the Authoring Version of the software, you will want to continue on to walk through construction of an interactive learning environment such as the one you played with at the outset of this tutorial.

5. Creating an Interactive Learning Environment
(Authoring Version Only)

As you saw in the first section of this tutorial, the Authoring Version of **STELLA II** allows you to create an interactive learning environment which will provide opportunities for others to create and experience the dynamics of the system you have modeled. In this section we will show you how to build an interface that will provide easy access to your model.

Open the model called "*Author*" found in the folder called "*Tutorial*." The model should open to the High-level Mapping and I/O Layer You should see three blank process frames. This is the model we used to create the first section of this tutorial.

stablishing Context In order for a person to get started with a model, we've found it useful to start them off with a high-level, big picture view. To do this, we recommend you have them come first to the High-level Map in order to establish a sense for the key "actors" in the system you have modeled. There are several options for establishing this big picture context for the user. These options are available by double-clicking within the process frame. When you do, the following dialog will appear:

At the top of the dialog, you can name the frame (and the underlying sector) and include documentation describing the assumptions in your model (this documentation will pop up when the end-users click in the "?" button in the process frame header). You can also import a picture (or movies if you are on the

Macintosh platform and have QuickTime) and change the border style. Feel free to edit the name, add some documentation, and change the border style. When you click OK you will see the changes on the map.

You have three choices for changing the face of any process frame or sector in your model. You can format the frame name and choose to have it appear in the process frame face as well as in the header. You can import a static picture or graphic. Finally, you can assign a QuickTime movie to play in the frame's face (Macintosh only).

Let's use the process frame name. Click on the button next to "Use Sector Name" and then click on the "Import Picture" button. You will get a dialog box which will allow you pick font and style for the lettering. Choose the options you want. When you click OK, you will see the changes on the map.

Repeat these steps for each of your process frames. Then we will develop the interactive portion of your interface.

Input Devices Scroll down the page until you have a blank screen in front of you. The tools you have available for giving the user input access to the model are the slider and the graphical function input and display device. For letting users see model output, you can create a graph, a table or a one-celled numeric display. The final device available is the "message poster," which will allow you to send messages to the user when certain criteria are met during the simulation. Let's start by putting some input devices on the screen.

Slider
Input Device Click once on the slider object (⊞) and slide the cursor into the screen. Click to deposit it where you want it. Add four more sliders (remember that you can hold down the alt key (Windows) or the option key (Macintosh) to retain the slider object). Double-click on the first one and the dialog below will appear:

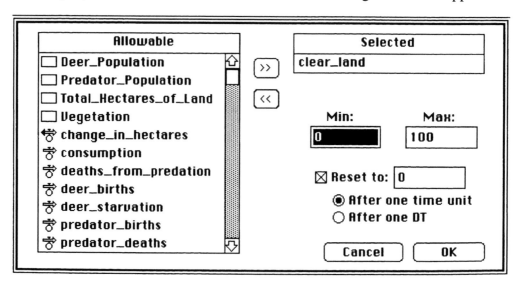

This dialog contains a list of all the variables in the model. The slider can be used to allow the users to set the initial value of a stock. It can also be used to set the constant value of a converter or a flow. You can even have the slider over-ride equation logic you have set up in the model.

Let's associate the first slider with the *clear land* converter (you will have to scroll down the list to find it). Move *clear land* over to the Selected list by double-clicking on the variable name or by selecting it and then clicking the >> button (similar to how you defined graphs and tables). You can only select one variable per slider.

You have two other options in this dialog: set the Min and Max, and Reset the value. When you set the Min and Max, you set limits for end-users preventing someone from picking an absurd value for the variable. Let's set a Min of 0 and a Max of 100 (hectares of land). We can also have the value reset to a predetermined value after one time unit, one dt or at the end of the run. In this case, we want *clear land* to be a one-time event that lasts for one year and then shuts off until the user selects it again. Click in the box next to "Reset to" and and type in 0. Leave the "After one time unit" choice selected.

Repeat the defining process for each of the other four sliders using these model variables: *deer tags*, *predator bounty*, *reintroduce deer* and *reintroduce predator*. You will want to have *reintroduce deer* and *reintroduce predator* reset to 0 in the same way that *clear land* does but the other two inputs should hold their value until the user changes them.

Graphical Function Input & Display Device (GFID)

The other input option available to us is the graphical function input and display device (GFID). Click once on the GFID ( ) and click again to place it on the control panel near the sliders. Double-click to open the associated dialog. It looks very similar to the slider dialog except the only entities listed are the graphical functions in the model. Choose *regeneration time*. Click OK. You are now back to the control panel. In order to change the shape of the curve, you (or your user) can double-click on the graphical function display icon and you will access the dialog seen at the top of the next page.

Here the end-users can change the relationship depicted in the graphical function without *permanently* changing it in the underlying model. If you want to assign this device to another graphical function, click on the "Delete Graph" button at the lower edge of the dialog box. This will return you to the define dialog to choose the new graphical function.

Documentation Caches

All sliders and GFID devices have the potential to have a "?" on their face. This is so you can provide on-line help to the user if they are uncertain what to do with one of the input devices on the screen. Let's put some documentation in the *clear land* slider to aid the user in understanding the purpose of the slider. Click on the downward-pointing arrow on the face of the *clear land* slider. You have

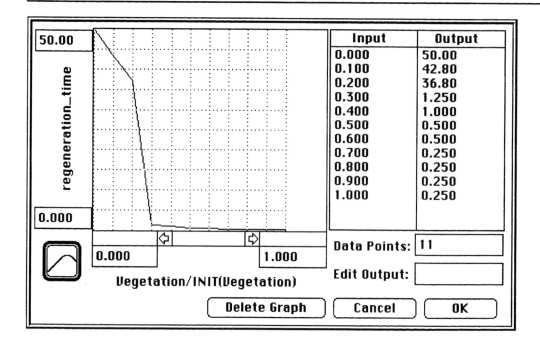

	Input	Output
50.00	0.000	50.00
	0.100	42.80
	0.200	36.80
	0.300	1.250
	0.400	1.000
	0.500	0.500
	0.600	0.500
	0.700	0.250
	0.800	0.250
	0.900	0.250
0.000	1.000	0.250

regeneration_time

0.000   **1.000**

Vegetation/INIT(Vegetation)

Data Points: 11

Edit Output:

[Delete Graph]   [Cancel]   [OK]

navigated to the *clear land* converter in the model (it is highlighted for you). Double-click on the *clear land* converter and click once on the Document button along the lower edge of the dialog box. You now see a documentation cache which will allow you type any information you want to include. When you finish typing your message, click OK. Then, click the upward-pointing arrow above the X^2 icon on the left border of the screen. Alternately, you can click the upward pointing arrow on one of the sectors. You are now back up on the High-level. Notice that a ? has appeared on the slider. If you click once on the ?, the text you just typed will pop up.

*Output
Devices*

Now you have provided end-users with the ability to give input to the model, but we need to provide them with the ability to see the resulting output as well. There are five choices for viewing model output on the High-level: graphs, tables, numeric displays, movies (Macintosh only) and GFID animation. The first two were introduced in the modeling section (Part 2) of this tutorial. Create a graph which displays the three major stocks in our model: *Deer Population*, *Vegetation*, *Predator Population*. **(If you have any problem doing so, refer to the Quick Help Guide that begins on page 88 of this booklet for details.)** Resize the graph (using the resize box in the lower right corner) and pin it (click once on the "stick-pin" in the top left corner) to the control panel. This allows the user to view model output and interact with the input devices at the same time (without the graph disappearing).

*Numeric
Display
Device*

Another useful output device is the numeric display device. This device provides you with the ability to show the user the current value of any variable without taking up the "real estate" on the control panel required by the table. Choose the numeric display device (⊂⊃) from the object palette and click it onto the control panel. Double-click on the device and you will get a dialog which allows you to choose the variable whose value you want to display. It works just like the slider and GFID. You may elect to have the device retain its ending value (a check box in the dialog). Otherwise, at the end of a run, the device will go blank. If you do choose "Retain Ending Value," the final value can be cleared under the Map & I/O Menu: select Restore, and then Numeric Display.

Message Posting   One way for you to give feedback to the end-user of your learning environment is through message posting. You can post a message when certain criteria is met in the model during the simulation run. This allows you to send warnings, give a pat on the back or draw their attention to some important dynamic in the model. This feature is accessed through any entity on the Diagram Layer of the model. A detailed description of Message Posting and how to implement it can be found in the *Authoring Module* in Chapter 6.

In addition to input and output devices, authored products can be enhanced by using the text block object and color. The text can give annotation on the control panel of your learning environment or section off parts, like we did on the PopDynam example in the first section of this tutorial. The frame and label for the Management Policies were done with a text block. Both of these were illustrated in Part 4 of this tutorial.

You now know enough about the Authoring features to build your own interactive learning environment. We encourage you to bring the insights and learning experiences you have had with the **STELLA II** software to the rest of the world using this interactive technology. Don't hesitate to contact us for technical support or for any additional help you need in developing your learning environment.

1. System Requirements:

Macintosh® version:
4mb RAM, Hard Disk with 5MB available, System 6.0.4 or higher
Fully System 7 compatible

Windows™ version:
4mb RAM, Hard Disk with 5MB available, 486 class macine running in the
386 enhanced mode, Windows version 3.1

2. Overview of STELLA II Operating Environment

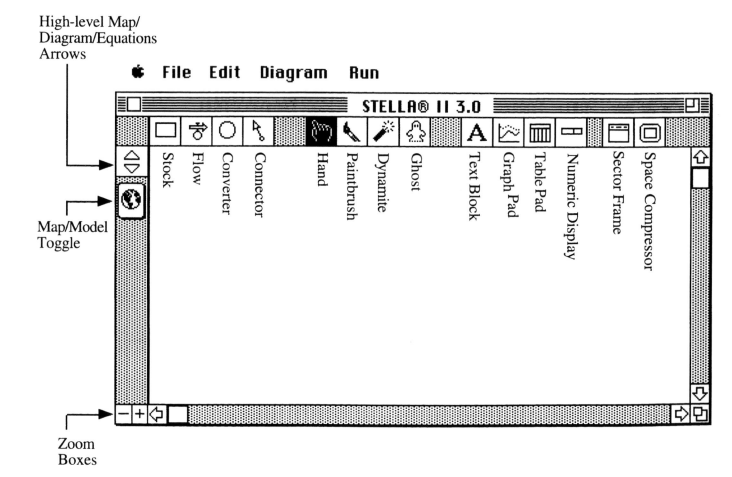

High-level Map/
Diagram/Equations
Arrows

Map/Model
Toggle

Zoom
Boxes

3. Drawing an Inflow to a Stock

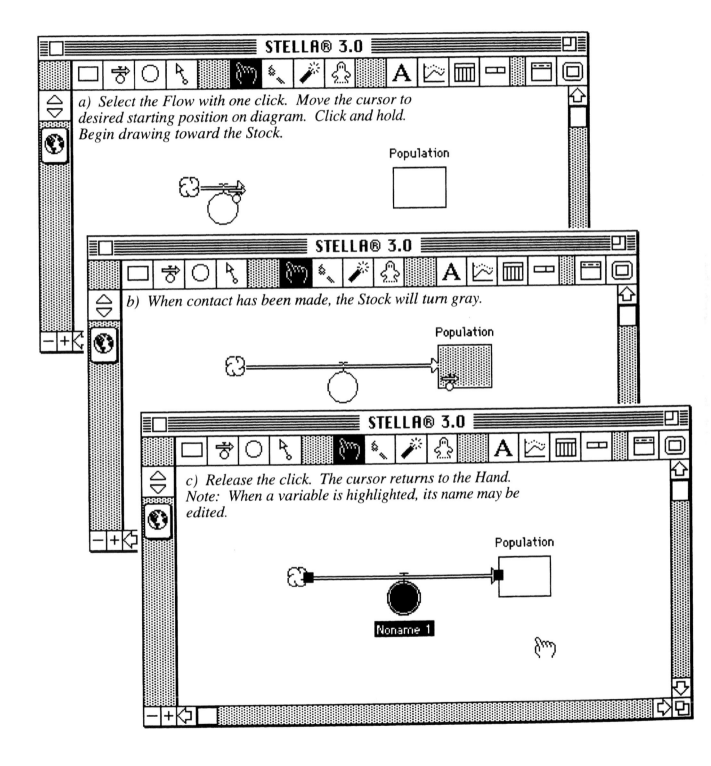

a) *Select the Flow with one click. Move the cursor to desired starting position on diagram. Click and hold. Begin drawing toward the Stock.*

Population

b) *When contact has been made, the Stock will turn gray.*

Population

c) *Release the click. The cursor returns to the Hand. Note: When a variable is highlighted, its name may be edited.*

Population

Noname 1

4. Drawing an Outflow from a Stock

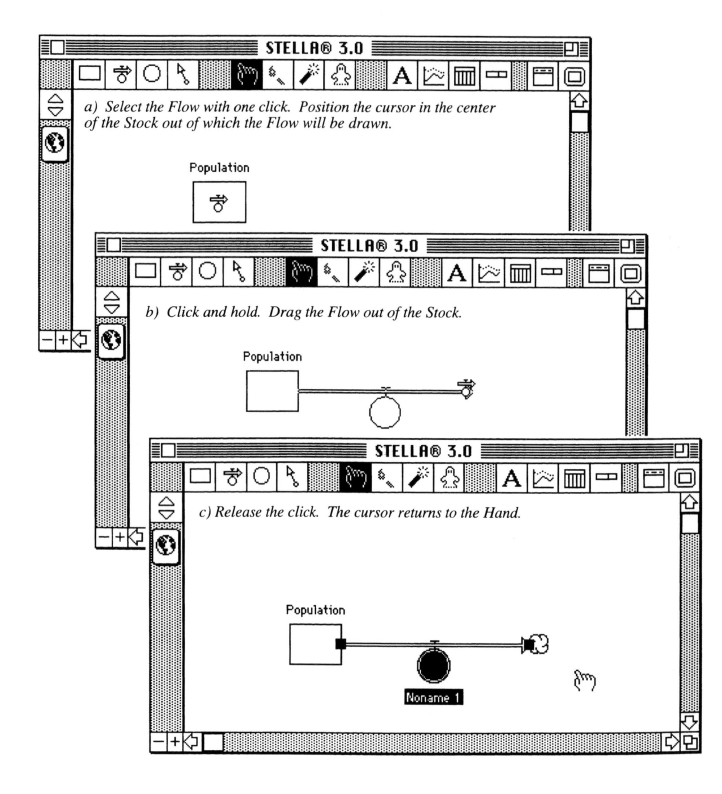

5. Replacing a Cloud with a Stock

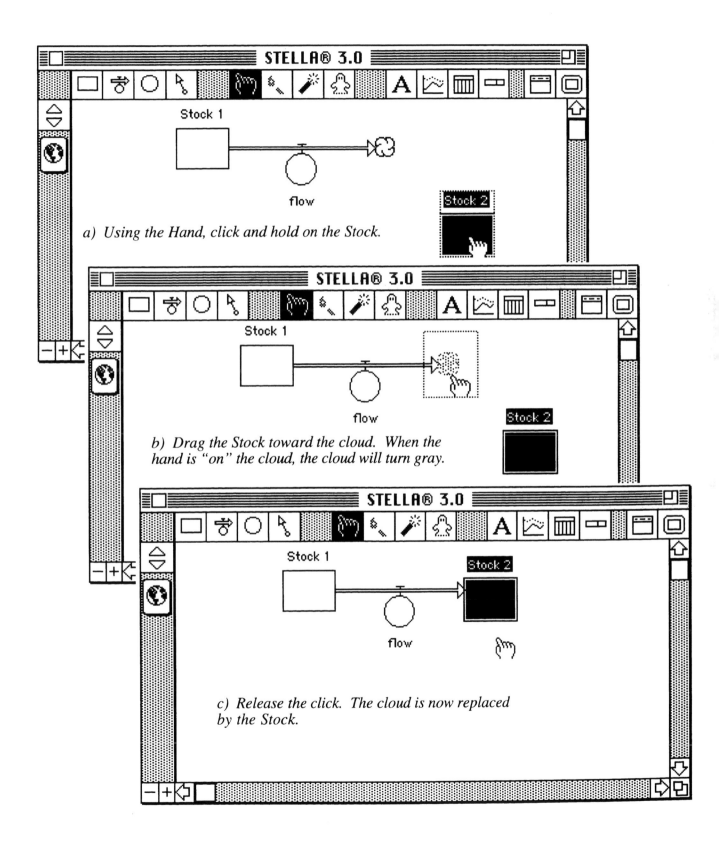

a) *Using the Hand, click and hold on the Stock.*

b) *Drag the Stock toward the cloud. When the hand is "on" the cloud, the cloud will turn gray.*

c) *Release the click. The cloud is now replaced by the Stock.*

6. Bending Flow Pipes

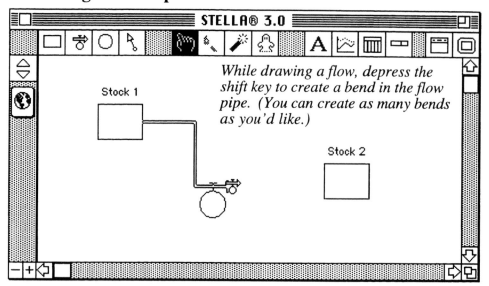

While drawing a flow, depress the shift key to create a bend in the flow pipe. (You can create as many bends as you'd like.)

7. Re-positioning Flow Pipes

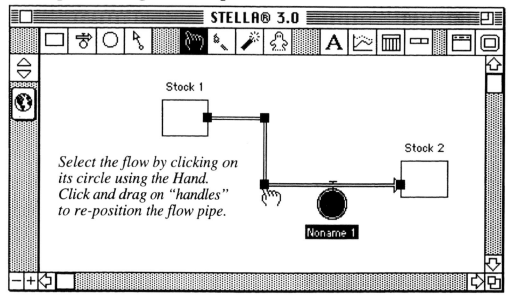

Select the flow by clicking on its circle using the Hand. Click and drag on "handles" to re-position the flow pipe.

8. Reversing direction of a Flow

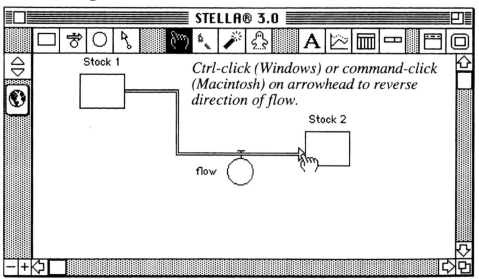

Ctrl-click (Windows) or command-click (Macintosh) on arrowhead to reverse direction of flow.

9. Flow Define Dialog; Builtins

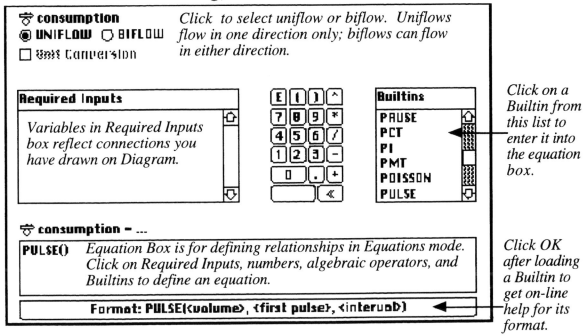

☡ consumption
◉ **UNIFLOW** ○ **BIFLOW**
☐ Unit Conversion

Click to select uniflow or biflow. Uniflows flow in one direction only; biflows can flow in either direction.

Required Inputs

Variables in Required Inputs box reflect connections you have drawn on Diagram.

E	(	)	^
7	8	9	*
4	5	6	/
1	2	3	–
0	.	+
	«

Builtins

PAUSE
PCT
PI
PMT
POISSON
PULSE

Click on a Builtin from this list to enter it into the equation box.

☡ consumption – ...

PULSE() *Equation Box is for defining relationships in Equations mode. Click on Required Inputs, numbers, algebraic operators, and Builtins to define an equation.*

Format: PULSE(<volume>, <first pulse>, <interval>)

Click OK after loading a Builtin to get on-line help for its format.

10. Moving variable names

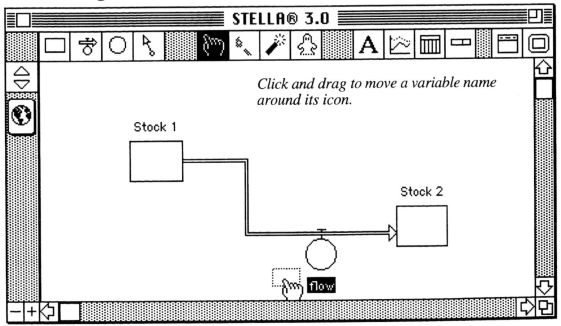

STELLA® 3.0

Click and drag to move a variable name around its icon.

Stock 1

Stock 2

flow

11. Drawing Connectors

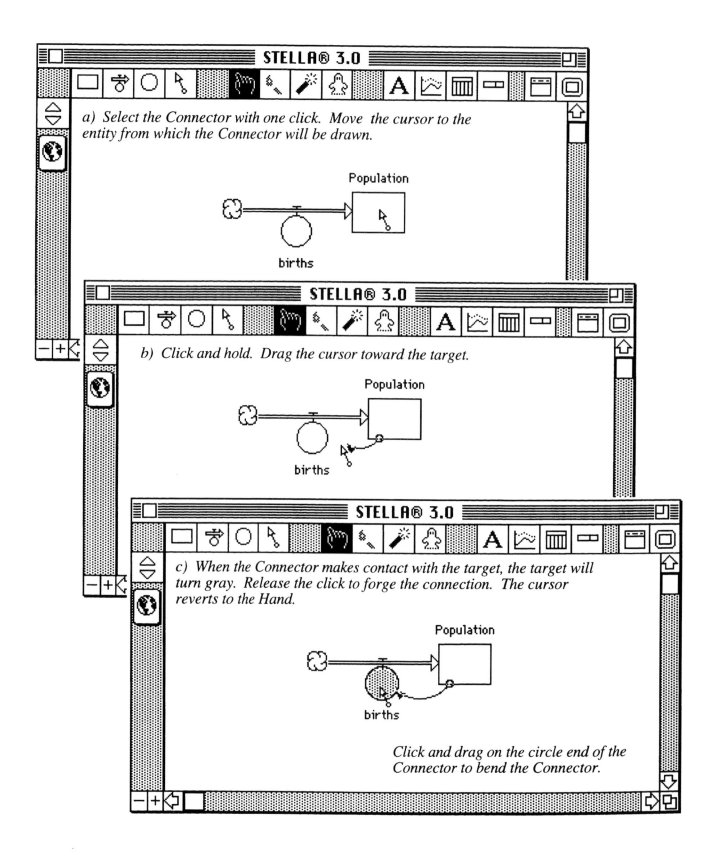

a) *Select the Connector with one click. Move the cursor to the entity from which the Connector will be drawn.*

Population

births

b) *Click and hold. Drag the cursor toward the target.*

Population

births

c) *When the Connector makes contact with the target, the target will turn gray. Release the click to forge the connection. The cursor reverts to the Hand.*

Population

births

Click and drag on the circle end of the Connector to bend the Connector.

12. Defining Graphs and Tables

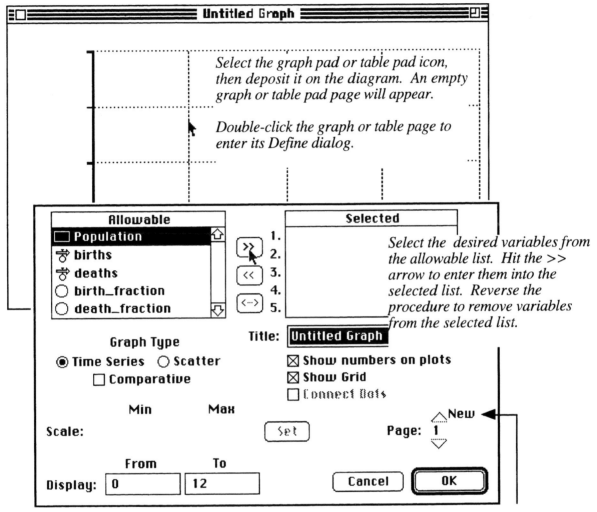

Select the graph pad or table pad icon, then deposit it on the diagram. An empty graph or table pad page will appear.

Double-click the graph or table page to enter its Define dialog.

Select the desired variables from the allowable list. Hit the >> arrow to enter them into the selected list. Reverse the procedure to remove variables from the selected list.

Note: Analogous operations for Table Pads.

Create as many new graph pad pages as you'd like.

13. Dynamite operations on Graphs and Tables

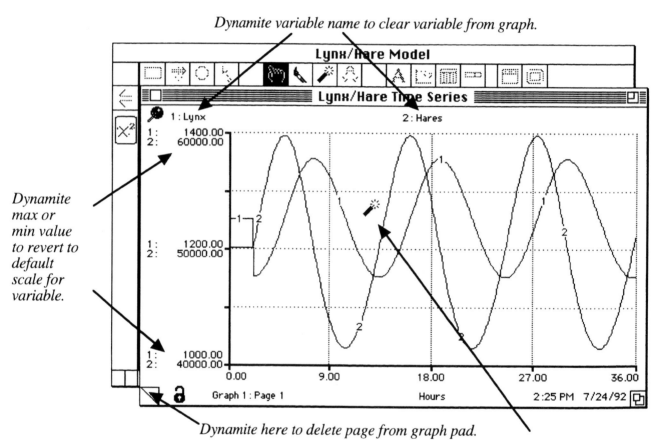

Dynamite variable name to clear variable from graph.

Dynamite max or min value to revert to default scale for variable.

Dynamite here to delete page from graph pad.

Dynamite here to clear data from graph pad page.

Note: Analogous operations for Table Pads.

INDEX